# Surface and Colloid Chemistry in Natural Waters and Water Treatment

# Surface and Colloid Chemistry in Natural Waters and Water Treatment

**Edited by**
## Ronald Beckett
Water Studies Centre
Monash University
Melbourne, Australia

**Plenum Press • New York and London**

Library of Congress Cataloging-in-Publication Data

---

Surface and colloid chemistry in natural waters and water treatment /
    edited by Ronald Beckett.
        p.    cm.
    "Proceedings based on a symposium on the role of surface and
colloid chemistry in natural waters and water treatment, held June
16-17, 1987, in Melbourne, Australia"--Verso ot t.p.
    Includes bibliographical references and index.
    ISBN 0-306-43802-X
    1. Surface chemistry--Congresses.  2. Colloids--Congresses.
3. Water chemistry--Congresses.    I. Beckett, Ronald.
QD506.A1S84    1990
546'.22--dc20                                           90-21580
                                                        CIP

---

Proceedings based on a symposium on The Role of Surface and Colloid
Chemistry in Natural Waters and Water Treatment, held June 16–17, 1987,
in Melbourne, Australia

ISBN 0-306-43802-X

© 1990 Plenum Press, New York
A Division of Plenum Publishing Corporation
233 Spring Street, New York, N.Y. 10013

Printed in the United States of America

# PREFACE

The discipline of surface and colloid chemistry has experienced a considerable resurgence since the early sixties. This perhaps reflects a growing realisation of the wide applicability of modern colloid and surface theory to many important industrial, medical and environmental problems. This increased activity has resulted in a very complex and at times even confusing area of science being consolidated within a firm theoretical framework.

The clearer insights gained into the underlying principles have no doubt acted in an autocatalytic manner to stimulate further interest in an expanding range of applications. A good example in the area of environmental chemistry has been the realization of the important role played by colloidal material and surface interactions in natural biogeochemical processes that has been the subject of increasing attention over the last few decades. This is well illustrated by the numerous studies carried out to elucidate the speciation, toxicity, transport and fate of pollutants in aquatic systems. In the vast majority of cases these have clearly implicated some involvement of an association between the pollutant (e.g. trace metal, toxic organic compound or nutrient) and a colloidal component (e.g. particle, humic substance, foam). In order to understand these interactions fully and their effect on pollutant mobility it is important to develop a full appreciation of the surface chemistry of these complex systems.

Australian scientists have long been prominent in the area of colloid and surface chemistry particularly during the latter half of this century. There now exists a strong network of academic departments and research groups around this country as well as in New Zealand and some of these have become involved in various aspects of aquatic colloid and surface chemistry.

Although not intended to be a comprehensive treatise, the aim of this book is to review the important role of surface and colloid chemistry in natural aquatic systems and in water treatment processes. The idea for the book developed following a two-day Symposium on the topic sponsored jointly by the Victorian Branch Physical Chemistry Group of the Royal Australian Chemical Institute and the Water Studies Centre, Department of Chemistry and Biology, Chisholm Institute of Technology, which was held in Melbourne on 16-17th June 1987. The volume gathers together chapters dealing with areas in which there have been significant advances in knowledge over the past few years. Each is written by a scientist from Australia or New Zealand who is actively involved in research into that particular field.

Some up to date reviews are included on topics of current interest such as humic substances, aquatic photochemistry, properties of natural microbes and water treatment processes. In addition the book contains some more detailed accounts of specific recent advances. Examples include the kinetics and mechanism of estuarine coagulation, transport of ground water radiocolloids, mechanism of water treatment processes involving magnetic particles and coagulation by iron salts, biological removal of phosphorous from wastewater and the effect of dam destratification on manganese speciation and behaviour.

The task of editing this book has been greatly facilitated by the quality of the contributions from my coauthors and their cooperation in performing the various editorial tasks so efficiently.

Finally, I wish to acknowledge Ms Rosa Villani who is largely responsible for the layout of the book and whose assistance in its production has been invaluable.

RON BECKETT
MELBOURNE
MARCH 1990

# CONTENTS

# SECTION I


# PROCESSES IN NATURAL WATERS

# THE SURFACE CHEMISTRY OF HUMIC SUBSTANCES

# IN AQUATIC SYSTEMS

Ronald Beckett

Water Studies Centre
Monash University
Melbourne, Victoria

## INTRODUCTION

The yellow, brown or red coloured organic substances that can be extracted from soils and sediments by alkaline solutions have long been recognised and studied (Achard, 1786; Saussure, 1804)   Their importance was probably first acknowledged by soil scientists in terms of their effect on soil condition (structure), retention of moisture and fertility and these workers carried out most of the early structure characterization (Stevenson, 1982).   In more recent years water scientists have recognised that similar compounds are present in almost all natural waters and that they also play an important role in many aquatic processes (Thurman, 1985; Aiken et al., 1985).

With the increased emphasis on environmental chemistry and also in the related discipline of fossil fuels, much research in the last decade or so has focussed on elucidating the structure of these ubiquitous humic substances.  The fact that humic substances are not simple compounds, but rather complex mixtures, is reflected by the rather vague definitions found in the literature.  Thus even in the excellent recent text by Thurman (1985), aquatic humic substances are defined simply as the "coloured, polyelectrolytic (organic) acids isolated from water by sorption onto (nonionic) XAD resins, weak-base anion exchange resins or a comparable procedure".  Such a definition is more a statement of how we can operationally isolate these compounds using affinity column chromatography and does not indicate precisely "what" they are.  Not only does their complexity hinder attempts at structure elucidation, but it probably means that the suite of molecules comprising the humic substances from one source is somewhat different from those collected from another environment.

Although commencing with a brief description of the nature of natural organic matter this review will concentrate on outlining some of the many processes occurring in natural waters in which humic substances play an important role.  As will be seen the majority of these are phenomena taking place at either the air/water or solid/water interfaces.  Humic substances play a dominant role in determining the colloid and surface properties of natural waters the importance of which will emerge in this and subsequent chapters of this book.

## THE ORIGIN AND NATURE OF HUMIC SUBSTANCES

Dissolved organic matter (DOM) appears to be present in all natural waters. Thurman (1985) has reviewed the available literature on the levels of dissolved organic matter in different waters.  The values found, usually expressed as dissolved organic

*Surface and Colloid Chemistry in Natural Waters and Water Treatment*
Edited by R. Beckett, Plenum Press, New York, 1990

carbon (DOC), vary enormously depending on the source of the water, geographic location and with season. River and lake waters can contain 1-60 mg/L of DOC although values between 2 and 10 mg/L are more typical. A combination of plant primary production and decomposition rates is usually the controlling factor in determining the DOC of a particular natural water. The levels of DOC in seawater, groundwater and rainwater is usually lower than in rivers and lakes and is generally less than 1 mg/L.

Usually between 40-90% of the total DOC in natural waters can be removed and isolated by the popular nonionic resin (e.g. XAD) adsorption procedure (Leenheer, 1981). In the majority of rivers these humic substances account for more like 40-60% of the DOC with higher proportions (70-90%) being found in highly coloured waters (Malcolm, 1985).

Humic substances can be subdivided into two fractions; humic acids which precipitate below pH 2 and fulvic acids which are soluble even at low pH. In most non-coloured waters the fulvic acid is the major fraction, typically making up about 90% of the humic substance or 45% of the total DOC present. In highly coloured waters the humic acid fraction can represent a much higher proportion of the DOC (Malcolm, 1985).

In general terms humic substances are yellow to brown in colour, and have elemental composition C = 45-55%, H = 4-5%, O = 35-40%, N = 1-2%, S and P < 1%. This suggests an unsaturated organic substance containing some conjugated chromophores with approximate average empirical formula $C_2H_2O$.

There has always been much conjecture concerning the exact molecular structure of humic substances. One view is that they consist of a complex mixture containing an almost infinite number of different structures perhaps suggesting that even to seek a structure elucidation is a futile task. Others venture to suggest specific structures or at least a typical molecule such as those depicted in Figure 1.

In the past humic substances have been thought of as being highly aromatic (Figure 1a), somewhat analogous to coal extracts, although more recent evidence, coming mainly from [13]C NMR studies (Wilson et al., 1987), suggests that perhaps only 20-40% of the carbons in aquatic humics are aromatic (Malcolm, 1985). Indeed this change in view is reflected quite well by the suggested structures depicted in Figures 1a through 1e. In general fulvic acids are found to be more aliphatic than humic acids with soil humic substances being more aromatic, but marine humic substances less aromatic in character, than those extracted from fresh waters (Thurman and Malcolm, 1983).

Humic substances are highly substituted with oxygen containing functional groups and these are responsible for many of their dominant properties such as, their water solubility, acidity, metal complexing capacity, surface activity and adsorption to particle surfaces. The major functional groups are carboxyl, phenol, hydroxyl, carbonyl, ether and ester and these are all represented in the structures in Figure 1.

There has been considerable controversy concerning the molecular weight of humic substances with values from 500 - 200 000 being quoted (Thurman, 1985). This may reflect certain limitations of the methods currently available and also perhaps the fact that these molecules appear to be capable of aggregation depending on the solution conditions.

The most reliable estimates are probably obtained from colligative properties, ultracentrifugation, low-angle X-ray scattering and flow field-flow fractionation. These techniques consistently yield an average molecular weight for fulvic acids of 800-1500 Dalton and for humic acids 1500-4000 Dalton (Beckett et al., 1987, 1989), which are somewhat lower than the values often quoted in the past. The molecular weight distributions of fulvic acids are narrower than those for humic acids as reflected in the weight average to number average molecular weight ratio ($Mw/Mn$) of 1.5-2 for fulvic acid, but up to 6 for certain humic acid samples. The average molecular weight for humic acids depends on their origin and has been found to increase in the order: **stream < soil < peat bog < coal** (Beckett et al., 1987).

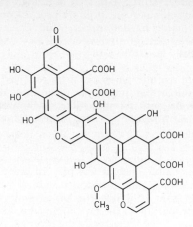

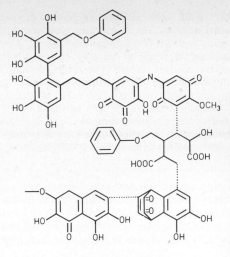

**A - Fuchs, 1930**

**B - Flaig, 1960**

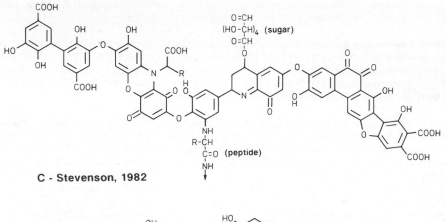

**C - Stevenson, 1982**

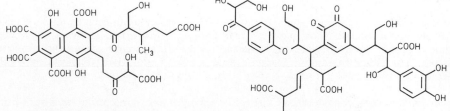

**D - Buffle, 1977**

**E - Steelink, 1985**

Figure 1. Structure of humic substances as proposed by various authors.

The exact mechanism involved in the formation of humic substances is not known, although it certainly involves microbial degradation of plant organic matter (e.g. lignins, cellulose, and polypeptides). In addition polymerization, condensation and oxidation reactions may occur modifying these degradation products. Aquatic humic substances may be derived by leaching of terrestrial plant and soil organic matter or may be formed by bacterial action on phytoplankton. The observed difference between aquatic humic substances and soil humic substances may be due to differential leaching of soil organic compounds, the inherent difference in the source of organic matter or may be due to ultraviolet radiation induced oxidation and polymerization that should have an increased opportunity of occurring in the organic enriched surface microlayer of natural waters (Harvey et al., 1983).

5

## Surface Activity

Surface activity is the pronounced tendency shown particularly by amphiphilic substances such as surfactants, to adsorb to air/water or solid/water interfaces. This phenomenon is promoted by molecules possessing distinct hydrophilic and hydrophobic portions. The opposing tendencies to dissolve in water can be accommodated by surface adsoption which in the case of the air/water interface is reflected by a decrease in the surface tension of the solution compared to that of pure water. The surface activity of some humic substances is demonstrated by the data given in Figure 2. The dashed curve shows the surface tension behaviour for a strong surfactant, sodium dodecylsulphate. This comparison indicates that the humic substances should be regarded as possessing moderate (or in some cases perhaps even strong) surface activity.

Lowering of the surface tension implies adsorption at the air/water interface as described by the Gibbs adsorption equation

$$\Gamma = \frac{-c}{RT}\frac{d\gamma}{dc} \quad mol\ m^{-2} \tag{1}$$

where

$\Gamma$ = surface excess concentration of solute (mol $m^{-2}$)
c = solution concentration of solute (mg/L or %)
$\gamma$ = surface tension (N $m^{-2}$)
R = gas constant (8.314 J $mol^{-1}$ $K^{-1}$)
T = absolute temperature (Kelvin)

Thus from a plot of $\gamma$ against c the surface excess concentration and thus the area occupied by a molecule at the air/water interface (in square Angstrom units) can be calculated at any given solution concentration

$$A = \frac{10^{20}}{\Gamma N_0} \quad A^2\ molecule^{-1} \tag{2}$$

where $N_0$ = Avogadro's number

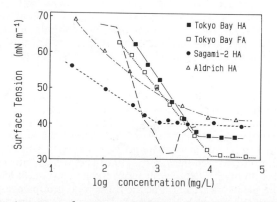

Figure 2.   Surface tension of some solutions of humic substances against concentration.   The humic samples were extracted from marine sediments (Tokyo Bay humic and fulvic acid from Hayase and Tsubota, 1983; Sagami-2 humic acid from Hayano et al., 1982) and a commercial humic acid obtained from Aldrich Co. Also plotted (dashed line) for comparison is data for a typical anionic surfactant (sodium dodecylsulphate (SDS)). The minimum is due to impurities.

The minimum area occupied by a molecule in a close packed surface layer has been estimated using this approach, with values ranging from 30-100 $A^2$ molecule[-1] being reported (Tschapek & Wasowski, 1976; Visser, 1982; Schnitzer, 1986). These values seem reasonable considering that the theoretical cross-sectional area of a -$CH_2$- group is about 22 $A^2$ and for the head group would be somewhat larger. These workers report that fulvic acids occupy a larger area per molecule than humic acids which could be due to increased steric or charge repulsion effects in the fulvic acids.

Visser (1982) has extended these calculations to obtain an estimated length/diameter aspect ratio for the humic molecules. These data indicate that the molecular configuration at the interface is rod-like with the elongation increasing with molecular weight. Whilst this is useful structural information it must be remembered that the conformation may well be different for molecules in free solution.

Using detailed viscosity measurements, Ghosh and Schnitzer (1980) deduced that under the solution conditions prevailing in waters and soils, humic substances behave like flexible linear colloids. However, they may coil up to give rigid spherocolloids at high solution concentration or in low pH and high ionic strength environments.

Ghosh and Schnitzer (1980) have suggested a method for molecular weight determination based on surface pressure measurements obtained using a Langmuir trough. However, this approach seems dubious as the equations used should only be applied in the case of insoluble monolayer formation at the interface.

Several trends in the surface activity of humic substances, as monitored by their effect on solution surface tension, have been noted by various authors. However, some of these do conflict. This may well be due, at least in part, to the pragmatic use of mass-based concentration (and other quantities), whereas a more revealing comparison could be obtained using molar quantities. The limitation here is the lack of reliable molecular weight data for the various humic samples investigated.

A number of workers have found that surface activity increases with average molecular weight of the sample (Hayano et al., 1982; Visser, 1982; Hayase and Tsubota, 1983). Yonebayashi and Hattori (1987) demonstrated that the larger molecular weight fraction of soil humic acids (i.e. the fraction virtually excluded in size exclusion chromatography) showed the greatest surface activity.

Visser (1982) found that aquatic humic substances are more surface active than terrestrial samples. He also reported that at low concentrations (say < 5 mg/L) aquatic humic acids are more surface active than fulvic acids, whereas at higher concentrations (> 25 mg/L) a reversal of this trend occurs. This latter observation is consistent with the data shown in Figure 2 for the Tokyo Bay marine sediment fulvic and humic acids (Hayase and Tsubota, 1983). We have also found that fulvic acids extracted from soils are much more surface active than the humic acid fraction (Beckett and Hoff, unpublished results).

Schnitzer (1986) reported that the surface tension of concentrated solutions (2-3%) of soil humic and fulvic acids decreased with increasing pH. In contrast, work in our laboratory (Beckett and Terrell, unpublished results) has shown that for Redwater Creek water concentrated to give a solution DOC of 400 mg/L, the surface tension decreased from about 70 mN m[-1] to 62 mN m[-1] as the pH was lowered from 9 to 3. Redwater Creek is a small highly coloured stream (situated in south western Victoria, Australia), with a natural DOC of about 25 mg/L. This latter trend indicates that, at least with this aquatic humic sample, as the pH is lowered and acidic functional groups are protonated, and the consequent increase in the hydrophobic nature of these groups provides the driving force causing an increase in the surface activity. Conversely with Schnitzer's soil humic substances presumably the major driving force is the existence of strongly hydrophilic groups which is promoted by hydrolysis of acidic moieties as the pH increases.

The obvious break in the surface tension versus concentration curves (see Figure 2) that occurs for some humic substances is suggestive of the micelle formation behaviour found in strong surfactants (Yonebayashi and Hattori, 1987). However, the occurrence of a limiting minimum surface tension is not sufficient to prove that micelles are formed in solution as it may simply represent the solution concentration at which the surface becomes saturated with adsorbed molecules. In any case this critical aggregation concentration, if indeed that is what it represents, invariably occurs at concentrations far in excess of the bulk solution concentrations of humic substances in natural waters. For example the break in the surface tension curves is typically found between 1-10 g/L, whereas natural levels of humics rarely exceed 0.1 g/L. However, concentrations high enough to promote aggregation may well exist in soil and sediment pore water and perhaps at the air/water surface microlayer, particularly in saline or low pH waters.

Wershaw (1986) argues strongly for the existence of aggregation of humic substances in soil and sediment pore waters. These aggregates could be in the form of micelles, membrane bilayers or vesicules. Wershaw et al (1986) propose similar models for the structure of humic materials in peat. The common feature of such structures is that they present a hydrophilic exterior to the aqueous phase and contain a hydrophobic (or lipophilic) interior region. Direct evidence for aggregation is obtained from low-angle X-ray scattering measurements and NMR spectral studies, and indirect evidence for these structures is inferred from the ability of humic substances to solubilize nonpolar solutes in water (see later section of this paper). Although this hypothesis is attractive, more evidence (e.g. as may be obtained from the direct electron microscopic observation methods used by Leppard et al., (1986)) is required to substantiate and refine the model structures proposed.

### The Surface Microlayer

As well as potentially lowering the surface tension of the solution, humic substances at the air/water interface may influence other processes such as exchange of gases between the atmosphere (Kanwisher, 1963; Hunter & Liss, 1981), dampening of surface waves (Blanchard, 1963; Garrett, 1967a), foam formation (Garrett, 1967b; Liss 1975; Hunter, 1980a) and concentration of particles and other elements and compounds at the surface (Hunter, 1980a). In addition to the Gibbsian monolayer it is important to consider the so called microlayer which is perhaps 100-500 µm thick depending on the method of collection and includes the hydrodynamic boundary layer of about 50 µm. This surface microlayer may contain up to 10 times the concentration of organic matter compared to the bulk water concentration (Hunter & Liss, 1981) and usually has elevated levels of trace metals (Wallace & Duce, 1975), toxic organics (Marty et al., 1978; Maguire et al., 1983) and microorganisms (Parker & Barsom, 1970). However, Hunter (1980a) has noted a depletion of Fe and Mn in a seawater microlayer presumably due to the dominance of rapid settling of larger mineral particles containing these elements.

Major processes contributing to the particulate flux into the microlayer include atmospheric settling and bubble flotation. A factor contributing to the elevated levels of particulate-associated elements in the microlayer is the increased residence time of particles at the surface caused by their tendency to adhere to the air/water interface. This occurs when the contact angle is large (Adamson, 1967), which Hunter (1980a) attributes to adsorption of natural organic matter lowering the surface energy compared to the oxide mineral surface, which would in turn lower the tendency for the particles to be wet by water.

The mechanisms discussed above are thought to result in increased loss of trace elements and particulate matter to the atmosphere by virtue of the elevated concentrations found in the surface microlayer and the ease with which aerosol droplets are produced mainly via bubble formation and bursting (Blanchard and Syzdek, 1974).

### HUMIC SUBSTANCES AT THE SOLID/WATER INTERFACE

Particulate matter, either in bottom sediments or suspended in the water column, forms an important compartment in any aquatic system (Beckett, 1986). It has a strong tendency to scavenge many types of pollutants (e.g. trace metals, toxic organics and nutrients). Thus, suspended particulate matter becomes important in determining the transport and fate of pollutants, and bottom sediments often become the ultimate sinks for these compounds.

Natural particulate matter is usually very heterogeneous in both chemical composition and particle size. It consists of a complex mixture of various minerals and biologically derived organic and inorganic material. However, a concept that has become widely accepted is that the surface layer of all these particles is coated with a common matrix consisting of hydrous metal oxides (e.g. iron, aluminium, silicon) and natural organic matter. Perhaps the most direct evidence for the existence of these surface coatings comes from the experiments outlined in the next section involving the measurement of surface charge or more correctly electrophoretic mobility.

Many processes occur at the particle/water interface, for example adsorption, desorption, precipitation, dissolution and aggregation. Although these processes are complex the surface coating model has proved to be a very useful simplifying assumption.

**Effect on Surface Charge**

The surface charge density of particles is an important quantity which influences a range of properties such as the colloid stability and adsorption behaviour. One of the most convenient techniques for monitoring the surface charge of dispersed particles is particle microelectrophoresis (Shaw, 1969). This measurement yields the electrophoretic mobility which is related, albeit in a fairly complex manner, to the zeta potential and surface charge density of the particles (Hunter, 1981). Studies of this type give considerable insight into the nature of the surface of natural aquatic particles.

The first significant finding from these experiments is that the mobility of the particles in a given sample fall in a fairly narrow range (Hunter and Liss, 1979), and with a few isolated exceptions (Newton and Liss, 1987) are always negatively charged. This result strongly suggests that all particles are covered with a common coating since the suspended particles should contain silica, clays, feldspars etc. which would be expected to be negatively charged, as well as iron and aluminium oxides which perhaps should be positively charged at natural pH (Stumm and Morgan, 1981). Hydrous oxides of iron, aluminium and perhaps manganese have been suggested as the coating material, but these would not be consistent with the high negative charge observed. A more consistent explanation is that the particles are coated with humic substances whose hydrolysable acidic functional groups could easily give rise to such a negative charge.

Supporting evidence for this humic coating hypothesis can be inferred from the fact that different colloids displaying a range of initial mobilities, all revert to a similar negative value after exposure to natural waters or dilute solutions of humic substances (Neihof and Loeb, 1972, 1974; Hunter, 1980b; Tipping, 1981a, 1981b; Tipping and Cooke, 1982; Loder and Liss, 1982, 1985). Thus, for example, we have found that a 20 mg/L goethite suspension with mobility $+3 \times 10^{-8}$ $m^2$ $s^{-1}$ $V^{-1}$ undergoes total charge reversal to a mobility of $-3 \times 10^{-8}$ $m^2$ $s^{-1}$ $V^{-1}$ on addition of less than 0.5 mg/L of humic or fulvic acid (Beckett and Le, 1990), which is far less than the concentration of humic substances found in most waters. This value of $-3 \times 10^{-8}$ $m^2$ $s^{-1}$ $V^{-1}$ represents a somewhat larger negative surface charge than found for natural suspended matter from fresh water systems. However, small amounts of divalent cations (e.g. $1 \times 10^{-4}$-$5 \times 10^{-4}$ M of Ca or Mg salts) will lower the magnitude of the surface charge density, so that the mobility becomes $-2 \times 10^{-8}$ $m^2$ $s^{-1}$ $V^{-1}$ or greater (i.e. lower magnitude). This now represents conditions very close to the situation found in many soft waters.

Several workers (Hunter and Liss, 1982; Beckett, 1986; Gerritsen and Bradley, 1987) have noted the inverse correlation between the magnitude of the mobility of natural particles and solution ionic strength (conductance) or more specifically the concentration of the so-called hard water cations Ca and Mg. This can be extended to higher ionic strengths to explain the surface charge behaviour of suspended particles in estuarie

(Pauc, 1980; Hunter and Liss, 1982; Beckett, 1986). Indeed Beckett and Le (1990) have demonstrated that the mobility trends of natural particles in freshwaters and estuaries can be modelled accurately using a synthetic goethite colloid whose surface has been saturated with a humic substance.

For example, Figure 3 shows the results obtained for the mobility in a series of mixing experiments involving either river suspended solids or humic-coated goethite with added Ca or Na salts. The river water particles and the humic-coated goethite particles behave very similarly. The data has been plotted on a salinity scale such that the abscissa at any point gives the salinity of a seawater-freshwater mixture that would contain the same concentration of the particular salt. The plots show very clearly that the lowering of particle mobility (magnitude) that occurs during seawater mixing in an estuary is due almost entirely to the presence of the minor divalent cations ($Ca^{2+}$ and $Mg^{2+}$), with the major ion in seawater ($Na^+$) having only a small influence on the mobility.

The above evidence points to the dominant role that adsorbed humic substances play in determining the surface charge of natural aquatic particles. In the case of positively charged particles (e.g. Fe and Al oxides), this no doubt involves the adsorption of negatively charged humic molecules although in addition to simple electrostatic attraction the mechanism may also involve covalent bond formation between surface OH groups and carboxylate functional groups on the humate ion (Tipping, 1981b). Tipping (1981a) has shown that the presence of the divalent cations, $Ca^{2+}$ and $Mg^{2+}$, increases the adsorption of humics onto iron oxides. This could either be due to the additional adsorption mechanism created by forming oxide-cation bridges or increases in the humic adsorption density due to cation binding of some of the active humate sites resulting in fewer humic-surface contacts per humic molecule being formed.

For clay minerals adsorption could occur at the positively charged edge planes. In addition Tombacz et al. (1988) have suggested that in the presence of sufficient electrolyte (e.g. for montmorillonite 0.2-0.5 M NaCl at pH 7) aggregation of the clay and humic substances may occur.

In the case of negatively charged particles, the bonding may involve a cationic intermediary. The ubiquitous iron and aluminium species are prime candidates for this role and it is conceivable that they could act either as bridging ions or by precipitation of a more strongly adsorbing oxide coating over the original mineral surface (James and Healy, 1972).

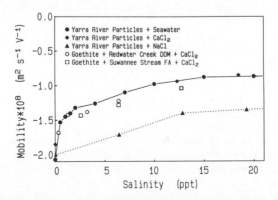

Figure 3. Electrophoretic mobility of Yarra River particles and a goethite colloid coated with humic substances from either Redwater Creek (Australia) or Suwannee River fulvic acid (USA) on mixing with various concentrations of seawater, $CaCl_2$ or NaCl. The abcissa is expressed in terms of the salinity (g/L) of a freshwater-seawater mixture that would contain the same concentration of the given salt.

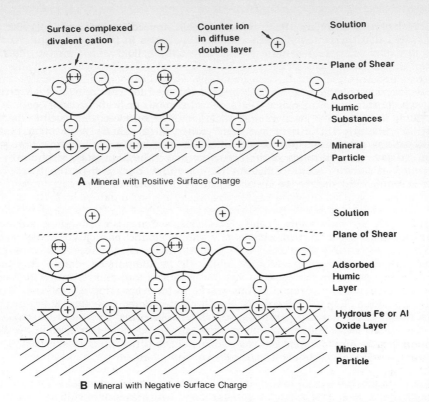

**A** Mineral with Positive Surface Charge

**B** Mineral with Negative Surface Charge

Figure 4. Schematic diagram depicting the role of adsorbed humic substances and divalent cations in modifying the surface charge of mineral particles.

These two conceptual models are depicted schematically in Figure 4. In either case the surface charge is controlled by the acidic functional groups on the outer film of adsorbed humic matter. This may be modified by complexation with suitable metal cations, particularly $Ca^{2+}$ and $Mg^{2+}$ ions which occur at various levels in natural waters.

## Effect On Aggregation

Colloidal particles in most freshwaters are quite stable to aggregation and the negative charge generated by the adsorbed film of humic substances is no doubt a major factor in maintaining the particles in a dispersed state. However, it is well known that aggregation of natural suspended particles (and also soils) can be induced by increasing the ionic strength. This most notably occurs during the mixing of freshwater and seawater in estuaries. It can also happen for inland waters by dissolution of calcareous rocks in hard water areas, by the influx of saline groundwater or in arid areas by evaporation.

In estuaries aggregation and settling is responsible for the nonconservative removal of suspended solids and associated elements from the water column (Boyle et al., 1974) and results in the deposition of large amounts of material to the bottom sediment. Major consequences of this process are the need to dredge most harbours to keep them open to shipping, the accumulation of high levels of pollutants in estuarine and near coastal sediments and the reduced flux of most elements to the open ocean.

Sholkovitz and coworkers have made an extensive study of estuarine coagulation of the filterable colloidal fraction in river water (e.g. Sholkovitz, 1976, 1978; Sholkovitz et al., 1978). An interesting finding was the fact that of the dominant divalent cations in seawater, Ca produces a greater extent of coagulation at a given concentration than does Mg. This demonstrates that as well as simple electrostatic effects, specific chemical interactions between the metal and colloid surface are involved in the coagulation mechanism (Eckert and Sholkovitz, 1976). Furthermore, Boyle et al. (1977) noted that the coagulation efficiency of divalent first row transition metal ions increased in the order Zn < Mn < Co < Ni < Cu which is the same order as their ability to form coordination complexes as expected from crystal field stabilization energy considerations (Irving and Williams, 1953). This perhaps reflects the fact that the coagulation mechanism may involve the formation of metal complexes with the functional groups contained in the organic molecules adsorbed to the surface of the particles.

There is considerable evidence to suggest that natural organic coatings inhibit the coagulation of natural particles. Gibbs (1983) showed that the rate of coagulation of sediment particles was increased considerably when their organic coatings were removed. Work conducted recently in our laboratory found that the colloidal iron in a highly coloured stream (Redwater Creek) with DOC   25 mg/L was much more resistant to coagulation by salts than the colloidal iron in a more typical river (Yarra) with DOC   8 mg/L (Nicholson, 1986). Several laboratory studies have shown that synthetic hydrous iron oxide colloids are stabilized by the addition of humic substances (Tipping and Higgins, 1982; Tipping, 1984, 1986; Cameron and Liss, 1984). In contrast Fox (1984) found that humic acid extracted from river water did not stabilize synthetic colloidal $Fe(OH)_3$, although he acknowledged that this could be due to the loss of much of the lower molecular weight fraction during the purification of the humic acid samples used.

Tipping and Ohnstad (1984a,b) conducted some detailed aggregation experiments comparing the behaviour of a natural iron oxide colloid collected from the metalimnion of a stratified lake with a synthetic humic-coated hematite colloid. In general, the coagulation behaviour of the two samples at the natural levels of particle and organic matter concentration was similar. This is consistent with the hypothesis that the colloid stability is controlled by an organic coating. At higher particle concentrations ($>2 \times 10^{-11}$ particles/L) and $CaCl_2$ concentrations (0.1 M), the coagulation rate of the natural particles was even greater than the maximum predicted for diffusion controlled aggregation. This accelerated kinetics was interpreted as evidence for the existence of bridging flocculation by organic matter analogous to polymer flocculation (La Mer, 1966). On the other hand, Tipping and Higgins (1982) deduced that in humic rich waters a steric stabilization contribution may also be present in addition to the usual electrostatic mechanism (Napper, 1985).

Jekel (1986) studied the effect of adsorbed humic substances on the coagulation behaviour of silica and kaolinite. A similar stabilization phenomenon was noted although the process is complicated by the fact that both mineral substrate and adsorbate are negatively charged. Both humic adsorption and colloid stabilization was promoted by lowering the solution pH and addition of small amounts of Ca (1 mM). It was deduced that adsorption was greatest for the humic fractions with lower charge density and higher molecular weight. Divalent cations probably influence the process by screening the negative charges on the solid surface and humic molecules thus enabling close approach of neutral parts of the molecule which leads to attachment. Alternatively, it is possible that $Ca^{2+}$ forms a bridge between the anionic groups on the surface and the organic compounds.

Although the adsorption of humic substances to silica and aluminosilicate minerals is inhibited by the negative surface charge developed at the pH of most natural waters (Davis and Gloor, 1981; Jekel, 1986), the surface of most natural particles is likely to be modified by the surface precipitation of hydrous iron or aluminium oxides. Since these have much higher isoelectric points (Stumm and Morgan, 1981) strong humic substance adsorption would be expected (see Figure 4b). It would seem that under the conditions prevailing in most natural waters the dominant effect of organic coatings on aggregation is to stabilize the particles against coagulation.

# EFFECT OF HUMIC SUBSTANCES ON POLLUTANT SPECIATION

There is ample evidence that humic substances associate with a variety of pollutant species either in solution or on the surface of suspended particulate matter. In some cases this binding involves the numerous polar or ionic functional groups (particularly those containing oxygen), whereas with hydrophobic solutes the presence of nonpolar regions of the humic molecules appears to be implicated. These interactions not only affect the solution speciation of the solute, but can also control the partitioning of pollutants between the aqueous and solid phases present. This is a rather complex process and humic-pollutant binding can in fact result in either increased or decreased adsorption to suspended particles. The forms of a pollutant have a major effect on its toxicity, transport and ultimate fate in the aquatic environment. The rates of uptake by biota, loss from the water column to the sediment and accumulation in the surface microlayer and hence possible loss to the atmosphere are all affected by the physico-chemical form. Thus, the determination of pollutant speciation has rightly received considerable attention by researchers, and a detailed understanding of the influence of man's activities on aquatic systems is dependent upon this knowledge.

## Trace metals

The fact that humic substances act as polyelectrolyte ligands for many trace metals in solution is well established and a quite voluminous literature has developed on the subject in recent years (see for example the articles and reviews by Mantoura & Riley, 1975; Buffle, et al., 1977, 1980; Mantoura et al, 1978; Hart, 1981; Alberts & Giesy, 1983; Neubecker & Allen, 1983; Kramer & Duniker, 1984; Boggs et al., 1985; Fish et al., 1986; Jones, 1987). The strength of these metal complexes varies with the metal and the functional group involved and are often moderately strong. For example, conditional stability constants for copper-humic complexes are often measured to be in the range $10^5$ - $10^9$ (Kramer & Duinker, 1984; Neubecker & Allen, 1983).

This complexation of metals will of course influence speciation which in turn affects the behaviour of these metals in natural water. For example, Daly (1986) reported that the toxicity of Cu to the freshwater shrimp *Paratya australiensis* depends mainly on the free $Cu^{2+}$ ion concentration. This is illustrated by the data in Table 1 which shows that the addition of waters containing complexing organic matter increases the 96 hour LC50 value whilst the measured $[Cu^{2+}]$ at the LC50 remains almost constant.

Complexing ligands can affect the adsorption of metals to particles in a number of ways. Theis and West (1986) demonstrated that cyanide can inhibit adsorption of Cu onto goethite by complexing the Cu and holding it in solution (Figure 5a). Davis and

Table 1.     96 Hour LC50 Values for *Paratya Australiensis* Exposed to Copper in the Presence of Natural Humic Substances (Daly, 1986)

| Water | DOC concentration (mg C/L) | Total Cu LC50 (µg Cu/L) | $[Cu^{2+}]$ present at LC50 (µg $Cu^{2+}$/L) |
|---|---|---|---|
| Melbourne Tap Water | 2 | 34 | 16 |
| Redwater Creek | 8 | 141 | 21 |
| Redwater Creek | 13 | 238 | 28 |

13

Leckie (1978) found a similar reduction in adsorption of Cu to ferric hydroxide by picolinic acid but attributed this effect to bonding of the picolinic acid to the surface and thereby blocking adsorption sites on the $Fe(OH)_3$ and limiting the Cu adsorption. On the other hand glutamic acid, apparently forms some sort of surface complex with Cu which enhances its adsorption (Figure 5b). Several other workers, for example Bourg et al. (1978, 1979) and Elliot and Huang (1979), have reported similar adsorption trends in various model systems.

Tipping et al. (1983) studied a ternary system containing freshwater humic substances, a synthetic goethite colloid and copper. They demonstrated that the humic-coated goethite showed an enhanced uptake of Cu compared to either the uncoated goethite or the humic material acting independently (Figure 5c). However, it should also be remembered that mobilisation of a metal from the adsorbed particulate phase can often be achieved by adding an excess of soluble complexing humic material to the suspension (Gleuck-Macholdt and Lieser, 1987). Certainly more evidence is needed to access the generality of these findings.

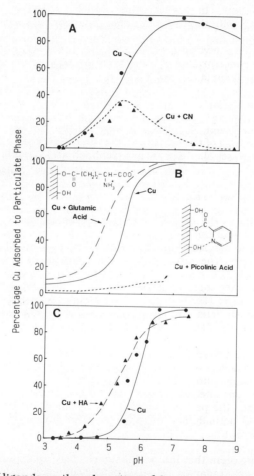

Figure 5. Effect of ligands on the adsorption of Cu onto iron oxides or hydroxide.
    (A) cyanide with goethite (Theis and West, 1986)
    (B) glutamic or picolinic acid with ferric hydroxide (Davis and Leckie, 1978)
    (C) humic acid with goethite (Tipping et al., 1983).

## Trace Organic Compounds

There is a considerable body of evidence pointing to the existence of an association between nonpolar organic molecules and humic substances. An indirect indication of this is the fact that much higher levels of many pesticides, herbicides and other toxic organics are found in sediments than in the water and the concentration can usually be correlated with the amount of organic matter in the sediment (Means et al., 1980; Means and Wijayaratne, 1982).

A somewhat more direct indication is the observation that the solubility of many of these sparingly soluble nonpolar compounds in water is increased significantly by the addition of humic substances (Chiou et al., 1986). The effectiveness of the dissolved organic matter to increase the solute solubility increases with its molecular size and hydrophobicity and for a given system the solubility increases linearly with the concentration of the added dissolved organic matter. These observations are consistent with a partition-like interaction (c.f. solvent extraction) of the solute with nonpolar microenvironments created by the humic substances.

It is possible that such microenvironments could be produced by association of humic molecules in solution as proposed in Wershaw's membrane model mentioned previously (Wershaw, 1986). In this case the increased solubility could be described as solubilization by analogy with the behaviour of micelle forming surfactants.

In adsorption studies involving soils and sediments it is found that the adsorption isotherms relating the concentration of sorbed solute ($[A]_{Ads}$) to the solution concentration ($[A]$) are usually linear (Chiou et al., 1979)

$$[A]_{Ads} = K_p[A] \tag{3}$$

The partition coefficient ($K_p$) is commonly related directly to the octanol-water distribution coefficient ($K_{ow}$) of the solute and the fractional organic carbon content ($f_{oc}$) of the sediment, (Karickhoff, et al., 1979; Schwarzenbach and Westall, 1981). Thus we can write

$$[A]_{Ads} = K_{ow} f_{oc} [A] \tag{4}$$

This is analogous to the behaviour of humic substances in solubilizing nonpolar solutes described above and can be used as evidence that the adsorption mechanism also involves partitioning into nonpolar regions created by the adsorbed organic matter (Chiou et al., 1979, 1983, 1985). The dependence of the sediment-water partition coefficient on the octanol-water distribution coefficient and apparent lack of sensitivity to the source of sedimentary organic matter suggests that the main driving force for the adsorption is the so-called hydrophobic effect (Tanford, 1980). Thus solute-solvent interactions appear to be of more importance in determining the affinity of these nonpolar compounds for the sediment than the strength of the solute-surface bond (Stumm, 1982; Chiou et al., 1983).

Recent studies conducted in our laboratories (Day et al., unpublished results) are aimed at assessing the effects of humic substances on the adsorption of several different herbicides. These studies compared the adsorption of each herbicide onto a synthetic goethite colloid either uncoated or coated with humic substances. The presence of a humic coating increased the partitioning of the nonpolar compound lindane by up to an order of magnitude but actually had the opposite effect on the adsorption of more polar compounds (e.g. glyphosate).

It would therefore seem that humic substances play an important role in the strong uptake of nonpolar toxic organics by suspended sediments in natural waters. These compounds are often very persistent and their ultimate fate and environmental effect will be determined to a large extent by the behaviour of the suspended particles. However, polar organics appear to be much more likely to remain in solution and in contrast these are often much more rapidly degraded by both chemical and biological processes.

## CONCLUSIONS

Humic substances appear to be present in all natural waters and play a vital role in determining many aquatic processes. They can be regarded as moderate surfactants and their presence lowers the surface tension of water and promotes foaming. They contribute to processes which result in elevated concentrations of various elements and particles in the so-called surface microlayer of waters. Humic substances coat the surface of aquatic particles ensuring they have a negative surface charge and increasing their colloid stability. Their ability to bind many substances will influence the speciation of pollutants such as trace metals and toxic organics. In particular they often increase the partitioning of pollutants into suspended particle and sediment phases in aquatic systems. In view of the importance of humic substances and their inherent structural complexity more work is certainly warranted in order to obtain a better mechanistic understanding of processes in natural waters.

## ACKNOWLEDGEMENTS

The parts of the work reported here which emanated from this laboratory was supported by the Australian Research Council. Several research students have been involved in this ongoing program in particular Geoff Day, Geoff Nicholson, Ngoc Le, Geoff Terrel and Solvi Hoff. I thank Barry Hart for some constructive suggestions on the manuscript.

## REFERENCES

Achard, F.K., 1786. Chemische Utersuchung des Torfs., *Crell's Chem. Ann.*, 2:391-403.

Adamson, A.W., 1967. "Physical Chemistry of Surfaces", 2nd Edition, Wiley-Interscience, New York.

Aiken, G.R., McKnight, D.M., Wershaw, R.L. and MacCarthy, P., 1985. "Humic Substances in Soil, Sediment and Water", Wiley-Interscience, New York.

Alberts, J.J. and Giesy, J.P., 1983. Conditional stability constants of trace metals and naturally occurring humic materials: application in equilibrium models and verification with field data, *in:* "Terrestrial and Aquatic Humic Materials", R.F. Christman, and E.T. Gjessing, eds., Ann Arbor Science, Ann Arbor, Michigan, pp 333-348.

Beckett, R., 1986. The composition and surface properties of suspended particulate matter, *in:* "The Role of Particulate Matter in the Transport and Fate of Pollutants" B.T. Hart, ed., Chisholm Institute of Technology, Melbourne, pp 113-141.

Beckett, R. and Le, N.P., 1990. The role of organic matter and ionic composition in determining the surface change of suspended particles in natural waters. *Coll. Surf.*, 44:35-49.

Beckett, R., Zhang, J. and Giddings, J.C., 1987. Determination of molecular weight distributions of fulvic and humic acids using flow field-flow fractionation, *Environ. Sci. Technol.*, 21:289-295.

Beckett, R, Bigelow, J.C., Zhang, J. and Giddings, J.C., 1989. Analysis of humic substances using flow field-flow fractionation, *in:* "The Influence of Aquatic Humic Substances on the Fate and Treatment of Pollutants", P. MacCarthy and I.H. Suffett, eds., ACS Advances in Chemistry Series No. 219, American Chemical Society, Washington, pp 65-80.

Blanchard, D.C., 1963. The electrification of the atmosphere by particles from bubbles in the sea, *Progr. Oceanogr.*, 1:73-202.

Blanchard, D.C. and Syzdek, L.D., 1974. Importance of bubble scavenging in the water to air transfer of organic material and bacteria, *J. Rech. Atmos.*, 8:529-540.

Boggs, S., Livermore, D.G. and Seitz, M.G., 1985. Humic macromolecules in natural waters, *JMS-Rev. Macromol. Chem. Phys.*, C25:599-657.

Bourg, A.C and Schindler, P.W., 1978. Ternary surface complexes. 1. Complex formation in the system silica-Cu(II)-ethylenediamine, *Chimia*, 32:166-169.

Bourg, A.C.M., Joss, S. and Schindler, P.W., 1979. Ternary surface complexes. 2. Complex formation in the system silica-Cu(II)-2,2'-bipyridyl, *Chimia*, 33:19-21.

Boyle, E., Collier, R., Dengler. A.T., Edmond, J.M., Ng, A.C. and Stallard, R.F., 1974. On the mass-balance in estuaries, *Geochim. Cosmochim. Acta*, 38:1719-1728.

Boyle, E.A., Edmond, J.M. and Sholkovitz, E.R., 1977. The mechanism of iron removal in estuaries, *Geochim. Cosmochim. Acta*, 41:1313-1324.

Buffle, J., 1977. Les substances humiques et leurs interactions avec les ions mineralaux, *in:* "Conference Proceedings de la commission d'Hydrologie Applique de l'A.G.H.T.M." L'Universite d'Orsay, pp 3-10.

Buffle, J., Greter, F.L. and Haerdi, W., 1977. Measurement of complexation properties of humic and fulvic acids in natural waters with lead and copper ion-selective electrodes, *Anal. Chem.*, 49:216-222.

Buffle, J., Deladoey, P., Greter, F.L. and Haerdi, W., 1980. Study of the complex formation of copper(II) by humic and fulvic substances, *Anal. Chim. Acta*, 116:255-274.

Cameron, A.J. and Liss, P.S., 1984. The Stabilization of "dissolved" iron in freshwaters, *Water Res.*, 18:179-185.

Chen, Y. and Schnitzer, M., 1978. The surface tension of aqueous solutions of soil humic substances, *Soil. Sci.*, 125:7-15.

Chiou, C.T., Peters, L.J. and Freed, V.H., 1979. A physical concept of soil-water equilibria for nonionic organic compounds, *Science*, 206:831-832.

Chiou, C T., Porter, P.E. and Schmedding, D.W., 1983. Partition equilibria of nonionic organic compounds between soil organic matter and water, *Environ. Sci. Technol.*, 17:227-231.

Chiou, C.T., Shoup, T.D. and Porter, P.E., 1985. Mechanistic roles of soil humus and minerals in the sorption of nonionic organic compounds from aqueous and organic solutions, *Org. Geochem.*, 8:9-14.

Chiou, C.T., Malcolm, R.L., Brinton, T.I. and Kile, D.E., 1986. Water solubility enhancement of some organic pollutants and pesticides by dissolved humic and fulvic acids, *Environ. Sci. Technol.*, 20:502-508.

Daly, H., 1986. An investigation of the relationship between copper speciation and toxicity to the freshwater shrimp *Paratya australiensis*", M. App. Sci. Thesis, Chisholm Institute of Technology, Melbourne.

Davis, J.A., 1978. Effect of adsorbed complexing ligands on trace metal uptake by hydrous oxides, *Environ. Sci. Technol.*, 12:1309-1315.

Davis, J.A. and Gloor, R., 1981. Adsorption of dissolved organics in lake water by aluminium oxide - effect of molecular weight, *Environ. Sci. Technol.*, 15:1223-1229.

Davis, J.A. and Leckie J., 1978. Effect of adsorbed complexing ligands on trace metal uptake by hydrous oxides, *Environ. Sci. Technol.*, 12:1309-1315.

Eckert, J.M. and Sholkovitz, E.R., 1976. The flocculation of iron, aluminium and humates from river water by electrolytes, *Geochim. Cosmochim. Acta.*, 40:847-848.

Elliot, H.A. and Huang, C., 1980. Adsorption of some copper(II)-amino acid complexes at the solid-solution interface. Effect of ligand and surface hydrophobicity, *Environ. Sci. Technol.*, 14:87-93.

Fish, W., Dzombak, D.A. and Morel, F.M.M., 1986. Metal-humate interactions. 2. application and comparison of models, *Environ. Sci. Technol.*, 20:676-683.

Flaig, W., 1960. Comparative chemical investigations on natural humic compounds and their model substances, *Sci. Proc. Roy. Dublin Soc.*, 4:49-62.

Fox, L.E., 1984. The relationship between dissolved humic acids and soluble iron in estuaries, *Geochim. Cosmochim. Acta*, 48:879-884.

Fuchs, W., 1930. Huminsauren, *Kolloid Z.*, 52:124-126.

Garrett, W.D., 1967a. Damping of capillary waves at the air-sea interface by oceanic surface active material, *J. Mar. Res.*, 25:279-291.

Garrett, W.D., 1967b. Stabilisation of air bubbles at the air-sea interface by surface active material, *Deep Sea Res.*, 14:661-672.

Gerritsen, J. and Bradley, S.W., 1987. Electrophoretic mobility of natural particles and cultured organisms in freshwaters, *Limnol. Oceanogr.*, 32:1049-1058.

Gibbs, R.J., 1983. The effect of natural organic coatings on the coagulation of particles, *Environ. Sci. Technol.*, 16:119-121.

Ghosh, K. and Schnitzer, M., 1980. Macromolecular structures of humic substances, *Soil Sci.*, 129:266-276.

Gleuck-Macholdt, C. and Lieser, K.H., 1987. Investigation of the behaviour of Zn, Cd, Hg and Pb in water without and with addition of suspended matter and complexing agents and in Rhine water by means of radionuclide techniques, *Vom Wasser*, 69:183-201.

Hart, B.T., 1981. Trace metal complexing capacity of natural waters: a review, *Environ. Technol. Lett.*, 2:95-110.

Harvey, G.R., Boran, D.A., Chesal, L.A. and Tokar, J.M., 1983. The structure of marine fulvic and humic acids, *Mar. Chem.*, 12:119-132.

Hayano, S., Shinozuka, N. and Hyakutake, M., 1982. Surface active properties of marine humic acids, *Yukagaku*, 31:357-362.

Hayase, K. and Tsubota, H., 1983. Sedimentary humic acid and fulvic acid surface active substances, *Geochim. Cosmochim. Acta*, 47:947-952.

Hunter, K.A., 1980a. Processes affecting particulate trace metals in the sea surface microlayer, *Mar. Chem.*, 9:49-70.

Hunter, K.A., 1980b. Microelectrophoretic properties of natural surface active organic matter in coastal seawater, *Limnol. Oceanogr.*, 25:807-822.

Hunter, R. J., 1981. "Zeta Potential in Colloid Science", Academic Press, London.

Hunter, K.A. and Liss, P.S., 1977. The input of organic material to the oceans: air-sea interactions and the organic chemical composition of the sea surface, *Mar. Chem.*, 5:361-379.

Hunter, K.A. and Liss, P.S., 1979. The surface charge of suspended particles in estuarine and coastal waters, *Nature*, 282:823-825.

Hunter, K.A. and Liss, P.S., 1981. Organic sea surface films, *in:* "Marine Organic Chemistry", E.K. Duursma and R. Dawson, eds., Elsevier, Amsterdam, pp259-298.

Hunter, K.A. and Liss, P.S., 1982. Organic matter and the surface charge of suspended particles in estuarine waters, *Limnol. Oceanogr.*, 27:322-335.

Irving, H.M. and Williams, R.J., 1953. Stability of transition metal complexes, *J. Chem. Soc. (London)*, 1953:3192.

James, R.O. and Healy, T.W., 1972. Adsorption of hydrolysable metal ions at the oxide water interface, *J. Coll. Interface Sci.*, 40:65-81.

Jekel, M.R., 1986. The stabilization of dispersed mineral particles by adsorption of humic substances, *Water Res.*, 20:1543-1554.

Jones, M.J., 1988. "The Development of Methods to Measure the Complexing Capacity of Natural Waters and their Application to Magela Creek Waters", M. App. Sci. Thesis, Chisholm Institute of Technology, Melbourne.

Kanwisher, J., 1963. On the exchange of gases between the atmosphere and the sea, *Deep Sea Res.*, 10:195-207.

Karickhoff, S.W., Brown, D.S. and Scott, T.A., 1979. Sorption of hydrophobic pollutants on natural sediments, *Water Res.*, 13:241-248.

Kramer, C.J.M. and Duinker, J.C., 1984. Complexation capacity and conditional stability constants for copper of sea and estuarine waters, sediment extracts and colloids *in:* "Complexation of Trace Metals by Natural Waters", C.J.M. Kramer and J.C. Duinker, ed., Nijhoff/Junk, The Hague, pp 217-228.

La Mer, V.K., 1966. Filtration of colloidal dispersions flocculated by anionic and cationic polyelectrolytes, *Disc. Farad. Soc.*, 42:248-254.

Leenheer, J.A., 1981. Comprehensive approach to preparative isolation and fractionation of dissolved organic carbon from natural waters and wastewaters, *Environ. Sci. Technol.*, 15:578-587.

Leppard, G.G., Buffle, J. and Baudat, R., 1986. A description of the aggregation properties of aquatic pedogenic fulvic acids, *Water. Res.*, 20:185-196.

Loder, T.C. and Liss, P.S., 1982. The role of organic matter in determining the surface charge of suspended particles in estuarine and oceanic waters, *Thalassia Jugoslavica*, 8:433-447.

Loder, T.C. and Liss, P.S., 1985. Control by organic coatings of the surface charge of estuarine suspended particles, *Limnol. Oceanogr.*, 30:418-421.

Maguire, R.J., Kurtz, K.W. and H G.G., Buffle, J. and Baudat, R., 1986. A description of the aggregation properties of aquatic pedogenic fulvic acids, *Water Res.*, 20:185-196.

Maguire, R.J., Kurtz, K.W. and Hale, E.J., 1983. Chlorinated hydrocarbons in the surface microlayer of the Niagra River, *J. Great Lakes Res.*, 9:281-286.

Malcolm, R.L., 1985. Humic substances in rivers and streams, *in:* "Humic Substances in Soil, Sediment and Water", G.R. Aiken, D.M. McKnight, R.L. Wershaw, and P. MacCarthy, eds., Wiley-Interscience, New York, pp 181-210.

Mantoura, R.F.C. and Riley, J.P., 1975. The use of gel filtration in the study of metal binding by humic acids and related compounds, *Anal. Chim. Acta*, 76:193-200.

Mantoura, R.F.C., Dickson, A. and Riley, J.P., 1978. The complexation of metals with humic materials in natural waters, *Estuarine Coast. Mar. Sci.*, 6:387-408.

Marty, J.C., Saliota, A. and Tissier, M.J., 1978. Hydrocarbons aliphatiques et polyaromatiques dans l'eau, la microcouche de surface et les aerosols marins en antlantique tropical est., *C.R. Acad. Sci. Paris, Ser. D*, 286:833-836.

Means, J.C. and Wijayaratne, R., 1982. Role of natural colloids in the transport of hydrophobic pollutants, *Science*, 215:968-970.

Means, J.C., Wood, S.G., Hassett, J.J. and Banwari, W.L., 1980. Sorption of polynuclear aromatic hydrocarbons by sediments and soils, *Environ. Sci. Technol.*, 14:1524-1528.

Neihof, R.A. and Loeb, G.I., 1972. The surface charge of particulate matter in seawater, *Limnol. Oceanogr.*, 17:7-16.

Neihof, R.A. and Loeb, G.I., 1974. Dissolved organic matter in seawater and the electric charge of immersed surfaces, *J. Mar. Res.*, 32:5-12.

Neubecker, T.A. and Allen, H.E., 1983. The measurement of complexation capacity and conditional stability constants for ligands in natural waters - a review, *Water Res.*, 17:1-14.

Newton, P.P. and Liss, P.S., 1987. Positively charged suspended particles: studies in an iron-rich river and its estuary, *Limnol. Oceanogr.*, 32:1267-1276.

Nicholson, G.J., 1987. "Studies on Sedimentation Field-Flow Fractionation and Coagulation of Fluvial Colloidal Matter", M. App. Sci. Thesis, Chisholm Institute of Technology, Melbourne.

Parker, B., and Barsom, G., 1970. Biological and chemical significance of surface microlayers in aquatic ecosystems, *Bioscience*, 20:87-92.

Pauc, H., 1980. Floculation et potentiel de surface des materiaux en suspension en environnement d'embouchure, *C. R. Acad. Sci. Paris, Series B*, 290:175-178.

Saussure, T.de, 1804. Recherches chemiques sur la vegetation, *Paris Ann.*, 12:162.

Schnitzer, M., 1986. Water retention by humic substances, *in:* "Peat and Water", C.H. Fuchsman, ed., Elsevier, New York, pp 159-176.

Schwarzenbach, R.P. and Westall, J., 1981. Transport of nonpolar organic compounds from surface water to groundwater, *Environ. Sci. Technol.*, 15:1360-1367.

Shaw, D.J., 1969. "Electrophoresis", Academic Press, New York.

Sholkovitz, E.R., 1976. Flocculation of dissolved organic and inorganic matter during the mixing of river water and seawater, *Geochim. Cosmochim. Acta*, 40:831-845.

Sholkovitz, E.R., 1978. The flocculation of dissolved Fe, Mn, Al, Cu, Ni, Co and Cd during estuarine mixing, *Earth Planet. Sci. Lett.*, 41:77-86.

Sholkovitz, E.R., Boyle, E.A. and Price, N.B., 1978. The removal of dissolved humic acids and iron during estuarine mixing, *Earth Planet. Sci. Lett.*, 41:130-136.

Steelink, C., 1985. Implications of elemental characteristics of humic substances, *in:* "Humic Substances in Soil, Sediment and Water", G.R. Aiken, D.M. McKnight, R.L. Wershaw, and P. MacCarthy, eds., Wiley-Interscience, New York, pp 457-476.

Stevenson, F.J., 1982. "Humus Chemistry", John Wiley, New York.

Stumm, W., 1982. Surface chemical theory as an aid to predict the distribution and the fate of trace constituents and pollutants in the aquatic environment, *Water Sci. Technol.*, 14:481-491.

Stumm, W. and Morgan, J.J., 1981. "Aquatic Chemistry", Wiley-Interscience, New York.

Tanford, C., 1980. "The Hydrophobic Effect", Wiley-Interscience, New York.

Theis, T.L. and West, M.J., 1986. Effects of cyanide complexation on adsorption of trace metals at the surface of goethite, *Environ. Technol. Lett.*, 7:309-318.

Thurman, E.M., 1985. "Organic Geochemistry of Natural Waters", Nijhoff/Junk, Dordrecht, Netherlands.

Thurman, E.M. and Malcolm, R.L., 1983. Structural study of humic substances: new approaches and methods, *in:* "Aquatic and Terrestrial Humic Materials", R.F. Christman and E.T. Gjessing, eds., Ann Arbor Science, Ann Arbor, pp 1-23.

Tipping, E., 1981a. The adsorption of humic substances by iron oxides, *Geochim. Cosmochim. Acta*, 45:191-199.

Tipping, E., 1981b. Adsorption by goethite (α-FeOOH) of humic substances from three different lakes, *Chem. Geol.*, 33:81-89.

Tipping, E., 1984. Humic substances and the surface properties of iron oxides in freshwaters, *in:* "Transfer Processes in Cohesive Sediment Systems", W.R. Parker and D.J.J. Kinsman, eds., Plenum, London, pp 31-46.

Tipping, E., 1986. Some aspects of the interactions between particulate oxides and aquatic humic substances, *Mar. Chem.*, 18:161-169.

Tipping, E. and Cooke, D., 1982. The effects of adsorbed humic substances on the surface charge of goethite (α-FeOOH) in freshwaters, *Geochim. Cosmochim. Acta,* 46:75-80.

Tipping, E. and Higgins, D.C., 1982. The effect of adsorbed humic substances on the colloid stability of haematite particles, *Coll. Surf.*, 5:85-92.

Tipping, E. and Ohnstad, M., 1984a. Colloid stability of iron oxide particles from a freshwater lake, *Nature*, 308:266-268.

Tipping, E. and Ohnstad, M., 1984b. Aggregation of aquatic humic substances, *Chem. Geol.*, 44:349-357.

Tipping, E., Griffith, J.R. and Hilton, J., 1983. The effect of adsorbed humic substances on the uptake of copper(II) by goethite, *Croat. Chem. Acta,* 56:613-621.

Tombacz, E., Gilde, M., A'braham, I. and Szanto, F., 1988. Effect of electrolyte concentration on the interaction of humic acid and humate with montmorillonite, *Applied Clay Sci.*, 3:31-52.

Tschapek, M. and Wasowski, C., 1976. The surface activity of humic acid, *Geochim. Cosmochim. Acta*, 40:1343-1345.

Visser, S.A., 1982. Surface active phenomena by humic substances of aquatic origin, *Rev. Franc. Sciences de l'Eau*, 1:285-296.

Wershaw, R.L., 1986. A new model for humic materials and their interactions with hydrophobic organic chemicals in soil-water or sediment-water systems, *J. Contam. Hydrol.*, 1:29-45.

Wershaw, R.L., Thorn, K.A., Pinckney, D.J., MacCarthy, P., Rice, J.A., and Hemond, H.F., 1986. Application of the membrane model to the secondary structure of humic materials in peat, *in:* "Peat and Water", C. H. Fuchsman, ed, Elsevier, Amsterdam, pp 133-157.

Wilson, M.A., Vassallo, A.M., Perdue, E.M. and Reuter, J.H., 1987. Compositional and solid-state NMR study of humic and fulvic acid fractions of soil organic matter, *Anal. Chem.*, 59:551-558.

Yonebayashi, K. and Hattori, T., 1987. Surface active properties of soil humic acids, *Sci. Total Environ.*, 62:55-64.

# MICROBIAL PROCESSES OCCURRING

## AT SURFACES

Kevin C. Marshall

School of Microbiology
The University of New South Wales
Kensington, New South Wales

## INTRODUCTION

Aquatic bacteria rapidly colonize surfaces submerged in flowing water and their subsequent growth and extracellular polymer production leads to the formation of biofilms on the exposed surfaces. A biofilm is mainly composed of a variety of bacteria embedded in a matrix of extracellular polymeric substances (EPS) of bacterial origin. The development of extensive biofilms is of considerable nuisance value on ship and pipeline surfaces, because of the induction of turbulent flow adjacent to the biofilm, and in heat-exchange systems, where biofilms reduce the degree of heat transfer. On the other hand, enhanced biofilm development is a desirable feature in such industrial applications as trickling filters for waste water treatment, fixed-film fermenters and fluidized-bed reactors. Why are bacteria so adept at adhering to surfaces and why do they grow so vigorously following colonization of such surfaces?

## BACTERIA AS LIVING COLLOIDS

Bacteria range in size from about 0.2 μm to 2.0 or more μm in length or diameter, but most bacteria in natural habitats tend towards the smaller end of this scale (van Es and Meyer-Reil, 1982). Generally, bacteria form stable colloidal suspensions in water despite the fact that the size of the cells exceeds the normal upper limit assigned to colloidal particles. This results partly from mutual electrostatic repulsion between the negatively charged bacteria and partly from the fact that the density of the bacterial cells is only slightly greater than that of water. The electrokinetic properties of bacteria may be determined by microelectrophoresis of cells suspended in buffers of different pH. Despite the presence of positively-charged surface ionogenic groups on many bacteria, the net charge on bacterial cells at neutral pH values is strongly negative (Marshall, 1976). The surface ionogenic properties of bacteria result from the particular chemical constitution of the cell envelope (wall) of the bacterium in question as well as the properties of EPS produced by the organism.

Different bacteria exhibit varying degrees of cell surface hydrophobicity, and it is believed that this characteristic is important in determining the extent of adhesion of bacteria to solid surfaces (Rosenberg and Kjelleberg, 1986). The overall colloidal nature of bacterial cells provides some insights as to why they readily adhere to so-called inert surfaces.

*Surface and Colloid Chemistry in Natural Waters and Water Treatment*
Edited by R. Beckett, Plenum Press, New York, 1990

# CHEMISTRY OF THE BACTERIAL CELL ENVELOPE

Although most metabolic functions of bacteria are localized within the plasma membrane, it is the cell wall and extracellular components that are in contact with the external environment and, hence, play a vital role in bacterial adhesion and biofilm properties. Bacteria have the simplest of cellular structures (termed prokaryotic cells) yet the cell envelopes of these organisms are remarkably complex. Most bacterial cells possess a rigid cell wall that is responsible for the characteristic morphology of the cell and for its structural integrity under variable osmotic conditions.

In Gram-positive bacteria the rigid component of the wall, peptidoglycan, accounts for 30 to 50% of isolated cell walls. Other secondary cell wall polymers in these bacteria include teichoic acids, teichuronic acids, neutral polysaccharides and, occasionally, protein. The Gram-positive cell wall is relatively thick and simple in its ultrastructure (as revealed under the electron microscope). Gram-negative cell walls are structurally complex with the peptidoglycan layer constituting only 2 to 10% of the dry weight of the wall. In these bacteria the plasma membrane and peptidoglycan layer are separated by a space, the periplasmic space, that is the site of activity of a number of hydrolytic enzymes. Outside of the peptidoglycan is the outer membrane, that consists of phospholipid, lipopolysaccharide and protein. This outer membrane is unusual in that it acts as a permeability barrier to hydrophobic substances. This is due to the asymmetry of the outer membrane resulting from all of the lipopolysaccharide occurring on the outer side of the bilayer whereas the phospholipid is found on the inner portion. A clear picture of the complexity of bacterial cell wall structures is given in the account by Wicken (1985).

Extracellular components that may be involved in adhesion processes include flagella (used in cell motility), fimbriae (proteinaceous fibrils involved in specific adhesion to living tissues), sex pili (involved in cell-to-cell contacts that result in conjugation), and extracellular polymeric substances (EPS, a variety of polymers that may result in non-specific adhesion to various surfaces and the formation of the matrix of biofilms). In the case of most aquatic bacteria, the EPS is the component leading to bacterial adhesion and biofilm development (see Marshall, 1986).

## ADHESION MECHANISMS

The adhesion of bacteria to surfaces has been considered both in terms of adhesion forces (colloidal stability) and a surface energy viewpoint. These two approaches are not necessarily mutually exclusive, as Pethica (1980) has pointed out that the total change in surface tension of opposing surfaces can be determined from the interacting forces and that this feature is automatically built into the colloid stability theory.

### Adhesion Forces

As stated earlier, bacteria normally possess a net negative surface charge and, consequently, they should rapidly adhere to any positively-charged surface. In natural habitats, however, most solid surfaces possess or assume a net negative charge. How do bacteria adhere to such surfaces?

(a) **Long range forces**. Bacteria are transported in the aqueous phase by fluid dynamic forces. As they approach a surface they become subjected to repulsion forces (resulting from the interaction between clouds of cations attracted to both the bacterial and substratum surfaces = electrical double layer repulsion) and to attraction forces of the London - van der Waals type. This is the basis of the DLVO theory that predicts a strong repulsion at low electrolyte concentrations, where the electrical double layer thickness is greater, and a weak attraction minimum (the secondary minimum) at higher electrolyte concentrations, where the cloud of cations is compressed and the electrical double layer thickness is minimal. It has been suggested (Marshall et al., 1971) that reversible adhesion of bacteria to surfaces and the spinning behaviour of motile bacteria upon contact with surfaces results from attraction to such a secondary minimum.

Direct observations of the attraction of the non-motile *Achromobacter* R8 to glass surfaces (Marshall et al., 1971) revealed that the bacteria were attracted to the surface at high electrolyte concentrations but, as predicted by the DLVO theory, were repelled at low electrolyte concentrations ($<5\times10^{-4}$M NaCl). However, invoking the idea of attraction to the secondary minimum may be overly simplistic in view of the previously mentioned complexity of both the bacterial cell envelope and the EPS (Rutter and Vincent, 1980). It could be that the observed stage of reversible adhesion (Marshall et al., 1971) may result from other factors such as minimal bridging contacts to the surface by EPS (see next section).

Bacteria unable to firmly adhere to surfaces are able to scavenge fatty acids bound to the substratum surface, an observation that also suggests the bacteria are held near the surface by long range forces (Hermansson and Marshall, 1985).

**(b) Short range forces**. Irreversible adhesion of bacteria to surfaces is achieved by the process of polymer bridging whereby EPS binds the cell to the substratum surface. Short range forces acting on individual polymer fibrils at the surface include (a) chemical bonds, such as covalent, electrostatic, and hydrogen bonds, (b) dipole interactions, such as dipole-dipole (Keesom), dipole-induced dipole (Debye), and ion-dipole interactions, and (c) hydrophobic interactions (Tadros, 1980). Although the precise chemical nature of most EPS is poorly understood, it is likely that different combinations of these short range forces will be involved in the interaction of a particular EPS with a variety of substratum surfaces. Examination of ultrathin sections of bacteria attached to surfaces has confirmed the role of polymer bridging in the firm adhesion of bacteria to surfaces (Fletcher and Floodgate, 1973; Marshall and Cruickshank, 1973).

## Surface Energy Approach

Despite the early recognition of the importance of bacterial and substratum surface properties in adhesion processes, many conflicting results were reported because investigations tended to consider only one of these parameters in any study. More recently, it has been realized that it is necessary to take account of at least three properties simultaneously, i.e. the substratum surface free energy, the bacterial surface free energy and the liquid surface tension, in any thermodynamic model of adhesion (Gerson and Scheer, 1980; Absolom et al., 1983; Fletcher and Pringle, 1985). In terms of the thermodynamic model, bacterial adhesion is favoured when the process results in a decrease in free energy. Neglecting systems in which electric charges and specific lectin-receptor interactions are involved, the change in free energy is given by

$$G_{adh} = \gamma_{BS} - \gamma_{BL} - \gamma_{SL} \tag{1}$$

where $G_{adh}$ is the change in free energy associated with bacterial adhesion, and $\gamma_{BS}$, $\gamma_{BL}$ and $\gamma_{SL}$ are the bacterium-substratum, bacterium-liquid and substratum-liquid interfacial tensions, respectively. Values for these interfacial tensions can be derived from Young's equation as follows

$$\gamma_{SV} - \gamma_{SL} = \gamma_{LV}\cos\theta \tag{2}$$

where $\gamma_{SV}$, $\gamma_{SL}$ and $\gamma_{LV}$ are the substratum-vapour, substratum-liquid and liquid-vapour interfacial tensions, respectively, and $\theta$ is the contact angle of the liquid on the substratum. Of these values, only $\gamma_{LV}$ and $\theta$ are readily determined experimentally but $\gamma_{SV}$ and $\gamma_{SL}$ may be calculated using an equation-of-state approach (Neumann et al., 1980). Within certain limitations, experimentally determined adhesion of a selected group of bacteria corresponded to theoretical predictions based on this thermodynamic model (Absolom et al., 1983). Other factors must be taken into consideration, however, such as the effects of growth conditions on bacterial surface free energy (McEldowney and Fletcher, 1986), the modifying effects of substratum conditioning films resulting from prior molecular adsorption to surfaces (Baier, 1980), and the fact that bacteria with an extreme degree of hydrophobicity have not been tested using the thermodynamic model.

23

# SIGNIFICANCE OF SURFACE-BOUND SUBSTRATE

Most aquatic ecosystems are oligotrophic, with a substrate flux of from near zero to less than one mg C/litre/day (Poindexter, 1981). Bacteria requiring relatively large amounts of energy substrate for growth and metabolism (copiotrophic bacteria) starve under oligotrophic conditions. Such bacteria respond to starvation conditions by a rapid reduction both in cell size and in endogenous respiration (Morita, 1982). In addition, these small, starved cells generally exhibit a greater propensity to adhere to solid surfaces than the well-fed bacteria (Dawson et al., 1981).

When an intermittent input of energy substrate into an oligotrophic ecosystem occurs, such as by the death and lysis of living organisms, bacteria rapidly utilize the soluble, low molecular weight organics in the water column and, thus, help to maintain the system in an oligotrophic state. When the readily available substrate is all utilized, the bacteria again begin to starve.

## Surface Conditioning Films

Macromolecules and low molecular weight hydrophobic organic molecules tend to rapidly partition at natural or artificial substratum surfaces immersed in a water body. The adsorption of such molecules at surfaces leads to the formation of conditioning films that alter the surface charge (Neihof and Loeb, 1972) and surface free energy (Baier, 1980) of the substratum. The formation of surface conditioning films precedes the adhesion of bacteria to newly immersed surfaces and the organic molecules of the conditioning films constitute an enriched source of substrate for bacteria gaining access to the substratum surface from oligotrophic habitats.

## Scavenging of Surface-Bound Substrates

Bacteria effectively scavenge labelled fatty acid bound to glass surfaces in situations where the fatty acid was the only substrate available to the organisms (Kefford et al., 1982). Bacteria irreversibly adhering to surfaces metabolized the bound fatty acid with a high degree of efficiency, yet bacteria capable only of reversible adhesion were also able to scavenge the surface-bound substrate prior to returning to the aqueous phase to seek other sites of nutrient concentration (Kefford et al., 1982; Hermansson and Marshall, 1985).

The amounts of adsorbed organics on surfaces in static water systems would be very limited, and attached bacteria would gain only a very transient benefit from being associated with the substratum surface under such circumstances.

In flowing systems, on the other hand, there should be a continual replenishment of organics at the substratum surface that would provide sufficient energy substrate to support active bacterial growth.

## Growth of Starved Bacteria at Surfaces

Growth of starved cells of the marine *Vibrio* DW1 has been observed at a membrane surface in a microdialysis chamber where a very dilute medium was employed (Kjelleberg et al., 1982). The bacteria were unable to grow in the aqueous phase because of insufficient energy substrate in this medium. Growth of the small cells to normal size followed by cell multiplication at the membrane surface occurred because of a concentration of organic substrate at this surface from the continuous flow past of the dilute medium. When the bacterial population at the surface built up to a high enough level, the rate of accumulation of substrate from the medium was insufficient to maintain active growth and the bacteria gradually showed signs of starvation once more (i.e. the cells reverted to small starvation-survival forms). Growth of bacteria from stream populations on undefined substrates at surfaces has recently been described using image-analysis techniques (Lawrence and Caldwell, 1987).

We have demonstrated that starved bacteria are capable of cellular growth and reproduction where the only source of energy is a defined substrate bound to a surface (Power and Marshall, 1988). In this study, stearic acid bound to the membrane surface of the microdialysis chamber was the source of substrate. Growth and reproduction of bacteria were observed by phase-contrast microscopy combined with time-lapse video techniques. The reversibly adhering *Vibrio* MH3 exhibited cellular growth and began the division process near the surface, but the final separation of daughter cells occurred after the bacteria returned to the aqueous phase. This process obviously provides a mechanism whereby such bacteria can benefit from substrate bound to surfaces, but they readily return to the aqueous phase where the bacteria are free to seek other sites of nutrient concentration.

In contrast to *Vibrio* MH3, cells of *Pseudomonas* JD8 remained attached to the surface, grew to normal size and, on division, the daughter cells slowly migrated across the substratum surface (Power and Marshall, 1988). This migration process, and the later detachment of cells from the surface, was thought to be related to changes in the substratum surface free energy following utilization of the surface-bound, hydrophobic stearic acid.

With a continuous adsorption of organics from solution in flowing systems, sufficient energy substrate should be available to allow growth and reproduction in the primary colonizing bacteria and the subsequent development of multilayers of bacteria and their EPS to give rise to biofilms on the substratum surface.


## CONCLUSIONS

In this brief paper, I have attempted to show that, by treating bacteria as living colloidal particles, it is possible to explain some of the mechanisms whereby bacteria interact with and firmly adhere to a substratum surface. As living organisms, even starving bacteria can respond to the presence of organic molecules adsorbed to substratum surfaces by metabolizing the organics as energy substrates. If sufficient substrate is available, as in flowing oligotrophic systems, then the adhering starved bacteria are capable of cellular growth and reproduction. Differing patterns of growth at surfaces have been observed with different bacteria. The continual growth of bacteria at surfaces and their production of abundant quantities of EPS eventually lead to the development of biofilms, i.e. microbial fouling of the surfaces.


## REFERENCES

Absolom, D.R., Lamberti, F.V., Policova, Z., Zingg, W., van Oss, C.J. and Neumann, A., 1983. Surface thermodynamics of bacterial adhesion, *Appl. Environ. Microbiol.*, 46:90.

Baier, R.E., 1980. Substrata influences on the adhesion of microorganisms and their resultant new surface properties. in: "Adsorption of Microorganisms to Surfaces". G. Bitton and K.C. Marshall, eds., Wiley-Interscience, New York.

Dawson, M.P., Humphrey, B.A. and Marshall, K.C., 1981. Adhesion: a tactic in the survival strategy of a marine *Vibrio* during starvation, *Curr. Microbiol.*, 6:195.

Fletcher, M. and Floodgate, G.D., 1973. An electron-microscopic demonstration of an acidic polysaccharide involved in the adhesion of a marine bacterium to solid surfaces *J. Gen. Microbiol.*, 74:325.

Fletcher, M. and Pringle, J.H., 1985. The effect of surface free energy and medium surface tension on bacterial attachment to solid surfaces, *J. Coll. Interface Sci.* 104:5.

Gerson, D.F. and Scheer, D., 1980. Cell surface energy, contact angles and medium surface tension on bacterial attachment to solid surfaces, *Biochim. Biophys. Acta*, 602:506.

Hermansson, M. and Marshall, K.C. 1985. Utilization of surface localised substrate by non-adhesive marine bacteria, *Microbial. Ecol.*, 12:91.

Kefford, B., Kjelleberg, S. and Marshall, K.C., 1982. Bacterial scavenging: Utilization of fatty acids localized at a solid-liquid interface, *Arch. Microbiol.*, 133:257.

Kjelleberg, S., Humphrey, B.A. and Marshall, K.C., 1982. The effect of interfaces on small starved marine bacteria, *Appl. Environ. Microbiol.*, 43:1166.

Lawrence, J.R. and Caldwell, D.E., 1987. Behavior of bacterial stream populations within the hydrodynamic boundary layers of surface microenvironments, *Microbial Ecol.* 14:15.

Marshall, K.C., 1976. "Interfaces in Microbial Ecology", Harvard University Press, Cambridge, Massachussets.

Marshall, K.C., 1986. Adsorption and adhesion processes in microbial growth at interfaces, *Adv. Coll. Interface Sci.*, 25:59.

Marshall, K.C. and Cruickshank, R.H., 1973. Cell surface hydrophobicity and the orientation of certain bacteria at interfaces, *Arch. Mikrobiol.*, 91:29.

Marshall, K.C., Stout, R. and Mitchell, R., 1971. Mechanism of the initial events in the sorption of marine bacteria to surfaces, *J. Gen. Microbiol.*, 68:337.

McEldowney, S. and Fletcher, M., 1986. Effect of growth conditions and surface characteristics of aquatic bacteria on their attachment to solid surfaces, *J. Gen. Microbiol.*, 132:513.

Morita, R.Y., 1982. Starvation-survival of heterotrophs in the marine environment, in: "Advances in Microbial Ecology, Vol. 6", K.C. Marshall, Ed., Plenum, New York.

Neihof, R.A. and Loeb, G.I., 1972. The surface charge of particulate matter in seawater, *Limnol. Oceanogr.*, 17:7.

Neumann, A.W., Hum, O.S., Francis, D.W., Zingg, W. and van Oss, C.J., 1980. Kinetic and thermodynamic aspects of platelet adhesion from suspension to various substrates, *J. Biomed. Mater. Res.*, 14:499.

Pethica, B.A., 1980. Microbial and cell adhesion, in: "Microbial Adhesion to Surfaces", R.C.W. Berkeley, J.M. Lynch, J. Melling, P.R. Rutter and B. Vincent, eds., Ellis Horwood, Chichester.

Poindexter, J.S., 1981. Oligotrophy: fast and famine existence, in: "Advances in Microbial Ecology, Vol. 5", M. Alexander, ed., Plenum, New York.

Power, K. and Marshall, K.C., 1988. Cellular growth and reproduction of marine bacteria on surface-bound substrate, *Biofouling*, 1:163.

Rosenberg, M. and Kjelleberg, S., 1986. Hydrophobic interactions: Role in bacterial adhesion, in: "Advances in Microbial Ecology, Vol. 9", K.C. Marshall, ed., Plenum Press, New York.

Rutter, P. and Vincent B., 1980. The adhesion of microorganisms to surfaces: physico-chemical aspects, in: "Microbial Adhesion to Surfaces", R.C.W. Berkeley, J.M. Lynch, J. Melling, P.R. Rutter and B. Vincent, eds., Ellis Horwood, Chichester.

Tadros, Th.F., 1980. Particle-surface adhesion, in: "Microbial Adhesion to Surfaces", R.C.W. Berkeley, J.M. Lynch, J. Melling, P.R. Rutter and B. Vincent, eds., Ellis Horwood, Chichester.

van Es, F.B., and Meyer-Reil, L.-A., 1982. Biomass and metabolic activity of heterotrophic marine bacteria, in: "Advances in Microbial Ecology, Vol. 6", K.C. Marshall, ed., Plenum Press, New York.

Wicken, A.J., 1985. Bacterial cell walls and surfaces, in: Bacterial Adhesion: Mechanisms and Physiological Significance", D.C. Savage and M. Fletcher, eds., Plenum Press, New York.

# PHOTOCHEMISTRY OF COLLOIDS AND SURFACES

# IN NATURAL WATERS AND WATER TREATMENT

T. David Waite

Australian Nuclear Science and Technology Organisation
Lucas Heights Research Laboratories
Lucas Heights, New South Wales

## INTRODUCTION

Particles in the upper layers of marine and freshwaters and in atmospheric water droplets are exposed to solar radiation and absorb a significant amount of this radiation. Although a large portion of the radiation absorbed by naturally occurring particles is converted into thermal energy, it is now recognized that a wide variety of chemical transformations involving suspended particulate matter are initiated by absorption of sunlight (Zafiriou et al., 1984; Zepp and Wolfe, 1987). While abiotic heterogeneous photochemical transformations involving living organisms and particulate organic materials have been reported (Zepp and Wolfe, 1987), photoprocesses involving particles containing transition metal ions are particularly numerous and are the focus of discussion in this chapter.

Recent studies into such heterogeneous processes as the light assisted dissolution of naturally occurring oxides and minerals and the photochemical abiotic fixation of nitrogen have advanced our understanding of fundamental aspects of natural aquatic systems. It has also been demonstrated over the last decade that heterogeneous photochemical processes play a key role in the degradation of anthropogenic chemicals such as petroleum discharges, pesticides, herbicides, and domestic and industrial wastes in natural environments. Evidence that heterogeneous photoprocesses are an effective method of degrading xenobiotic compounds in aqueous solution has resulted in extensive investigation into the possibility of incorporating a photochemical stage in water and waste treatment plants.

The growing interest in aquatic photochemistry is evidenced by the number of papers and reviews in this field. In addition to the references given above, a number of reviews and compilations on the general topic of photoprocesses in natural waters are available (see, for example, Zepp, 1980; Zika, 1981; Zafiriou, 1983; Zika and Cooper, 1987; Zika, 1987). Specific reviews of investigations into aspects of photoprocesses at the particle-water interface of relevance to natural waters and waste treatment have been prepared by Waite (1986) and Langford and Carey (1987). Considerable advances have been made both in our understanding of mechanistic aspects of heterogeneous photoprocesses and in the application of these processes to water and waste treatment.

*Surface and Colloid Chemistry in Natural Waters and Water Treatment*
Edited by R. Beckett, Plenum Press, New York, 1990

## Dissolution of Naturally Occurring Oxide/Hydroxide Minerals

Oxide/hydroxide minerals of Mn(III,IV) and Fe(III) are thermodynamically stable under most oxygenated natural water conditions and typically exhibit high surface areas (particularly if in colloidal form). As such, these minerals represent ideal loci for sorption of a wide range of water constituents including potentially troublesome species such as phosphate and trace (often toxic) metals and organic compounds (including pesticides, herbicides and aromatic hydrocarbons). Changes in oxidation state dramatically alter the solubility of these oxides with consequent implications for the fate of any adsorbed species. Thus, reduction of Fe(III) to Fe(II) increases iron solubility with respect to oxide/hydroxide phases by as much as eight orders of magnitude (Stumm and Morgan, 1981). In addition to the implications for the fate of adsorbed toxic or troublesome species, such changes in the solubility of iron and manganese may have a significant effect on the biota for which these metal ions are essential nutrients but, typically, unavailable in their particulate form. Indeed, Anderson and Morel (1983), Finden et al. (1984), Sunda et al. (1983) and Sunda (1989) note that iron or manganese may limit algal productivity in some situations and speculate that in such cases, enhancement of limiting nutrient supply through reduction processes may significantly alter algal growth rate, primary productivity, and species distribution. Field evidence that iron or manganese are limiting nutrients in some situations has been provided by Brand et al. (1983), Entsch et al. (1983), Martin and Gordon (1988), and Martin and Fitzwater (1988).

A variety of field and laboratory studies over the last five years have indicated that solar radiation may induce or enhance the reductive dissolution of iron and manganese oxides/hydroxides in oxygenated natural waters. Collienne (1983) and McKnight and Bencala (1988) present convincing evidence for a sunlight-driven diurnal cycling in Fe(II) concentration in acidic streams and lakes which involves a concomitant dissolution of solid iron oxide/hydroxide (see, for example, Figure 1). Sunda et al. (1983) have shown that light enhances the dissolution of an amorphous manganese oxide in seawater and have suggested that this process accounts for the surface maxima in dissolved manganese observed in the oceans.

In most of the cases cited above, the presence of natural organic materials was shown to be critical to the occurrence of photodissolution. That naturally occurring organic compounds are capable of inducing or assisting the photodissolution of iron and

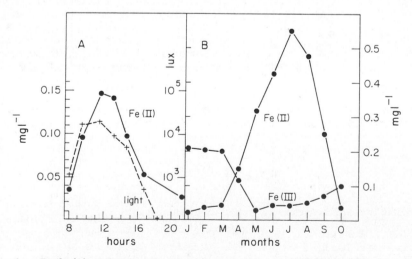

Figure 1.  Daily (A) and seasonal (B) accumulation of photochemically produced ferrous ion in a well-oxygenated acid lake (Reproduced from Morel, 1983; adapted from Collienne, 1983).

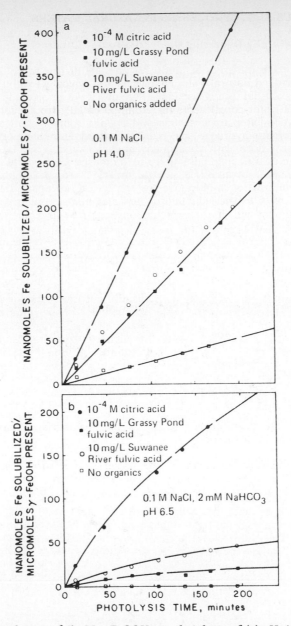

Figure 2. Dissolution of 5 μM γ-FeOOH on photolysis of (a) pH 4.0, and (b) pH 6.5 solutions containing either $10^{-4}$ M citrate, 10 mg/L aquatic fulvic acids, or no added organic agent. Light source: simulated AM1 solar spectrum (Reproduced from Waite and Morel, 1984a).

manganese oxides has been confirmed in a number of laboratory studies using humic and fulvic acids extracted from fresh and marine waters (Waite and Morel, 1984a; Sunda et al., 1983; Waite et al., 1988). As shown in Figure 2, aquatic fulvic acids (at concentrations of 10 mg/L) extracted from two different sources resulted in approximately four-fold increases in the rate of solar radiation-induced dissolution of 5 μM suspensions of the commonly occurring iron oxide phase lepidocrocite (γ-FeOOH) at pH 4.0 compared to that observed in an organic free suspension (the extent of dissolution in both the presence and absence of fulvic acids under dark conditions was negligible over the time scale of interest). At pH 6.5, photolysis increased the rate of dissolution from near zero to measurable rates. The effect of near-UV light (300-400 nm; $\lambda_{max} = 365$ nm) in enhancing the dissolution of 5 µM of the synthetic manganese oxide vernadite, $\delta$-$MnO_2$ in the presence of 10 mg/L Suwannee River fulvic acid (an International Humic Substances Society standard material) is somewhat more dramatic with approximately 90% solubilization after three hours of continuous photolysis compared with only 20% dissolution after suspension in the same medium for three hours with no illumination (Figure 3).

Considerable insight into the possible mechanisms of photodissolution of metal oxides/hydroxides in natural waters may be gleaned from the results of studies in simple, well-defined systems. As can be seen from Figure 2, photolysis of $\gamma$-FeOOH in the presence of the strong Fe(III) complexing agent citric acid results in dissolution phenomena qualitatively similar to that observed in the presence of fulvic acid. Waite and Morel (1984b) proposed three alternative mechanisms for the preliminary photochemical process in this latter case: (a) a photo-Kolbe process in which photo-generated holes are scavenged by readily oxidizable species such as the RCOO⁻ groups and the remaining electrons reduce the colloidal substrate (a semiconductor mechanism), (b) photo-degradation of surface-located Fe(III)-OH groups, assisted by ·OH scavenging by citrate, and (c) a ligand to metal charge transfer (LMCT) process within surface-located Fe(III)-citrate groups resulting in oxidation of the ligand and the reduction of the Fe(III) metal center (a process equivalent to the well-documented homogeneous photolysis of ferric citrate). Mechanism (b) was discarded because the surface-located Fe(III) hydroxo groups have been reported to absorb at considerably higher energy than that of the irradiating light used, but both (a) and (c) remain possibilities. All iron oxides/hydroxides absorb strongly in the near-UV region (Sherman and Waite, 1985) thus excitation by solar radiation may generate electron-hole pairs by promotion of an electron from the valence to the conduction band. This process is depicted schematically in Figure 4. While e⁻-h⁺ pair recombination appears to be rapid in iron oxides (Stramel and Thomas, 1986; Leland and Bard, 1987), hole transfer to sorbed species should be fast enough to ensure that some degree of oxidation takes place at the oxide surface. Electrons remaining at the surface after removal of holes have been shown to possess lifetimes of the order of milliseconds (Frese and Kennedy, 1983) and may reduce sorbed oxygen or may reductively dissolve the solid.

The photooxidation of oxalate, sulphite, and iodide by iron oxides with concomitant release of Fe(II) has been described by Leland and Bard (1987) in terms of a semiconductor mechanism although differences in the rates of photooxidation between the various iron oxide phases appeared to be due to intrinsic differences in crystal and surface structure rather than differences in surface area or band gap. Faust and

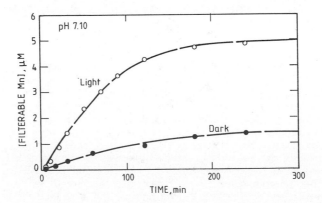

Figure 3.   Dissolution of 5 µM $\delta$-$MnO_2$ in 10 mg/L fulvic acid under dark and light conditions. Light source: 100 Watt Hg arc lamp with 365 nm narrow band-pass filter, (Adapted from Waite et al., 1988).

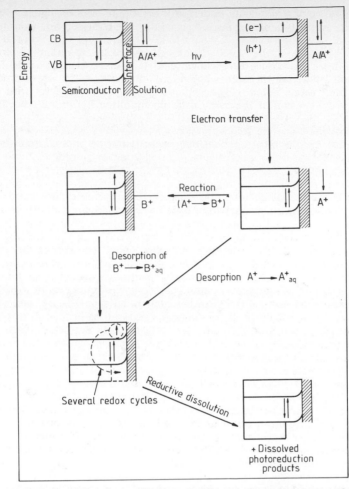

Figure 4.    One possible scheme for semiconductor photochemistry, where e⁻ and h⁺ are electrons and holes respectively and VB and CB represent the valence and conduction bands respectively. After creation of e⁻ and h⁺ pairs in the solid, recombination competes with diffusion to the surface, where interaction with adsorbed molecules may lead to trapping of e⁻ or h⁺ and desorption of the oxidized or reduced products. Scavenging of h⁺ may generate excess e⁻ in the solid, leading to reductive photodissolution of the crystal (Reproduced from Zafiriou et al., 1984).

Hoffmann (1986) also investigated the photodissolution of hematite in the presence of sulphite and observed a significant increase in the Fe(II)$_{aq}$ quantum yield at 367 nm, the peak wavelength of the aqueous phase Fe(III)-S(IV) complex LMCT band. Given that the LMCT band of any surface Fe(III)-S(IV) complex is likely to occur at a similar wavelength, photo-induced charge transfer within the surface-located complex provides the simplest explanation consistent with experimental results, but reductive dissolution as a consequence of direct excitation of the bulk solid cannot be discounted. A conceptual model of the iron redox chemistry in hematite suspensions containing S(IV) is shown in Figure 5. Litter and Blesa (1988) concluded that a semiconductor mechanism reasonably accounted for the photodissolution of maghemite ($\gamma$-Fe$_2$O$_3$) in the presence of ethylenediaminetetraacetic acid (EDTA) but found that EDTA oxidation also occurred only when surface-located EDTA-Fe(III) LMCT bands were excited. There is evidence also that, in some cases, solution phase photoprocesses may play an important role in dissolution of metal oxides/hydroxides. Thus, Cornell and Schindler (1987) observed an initial slow

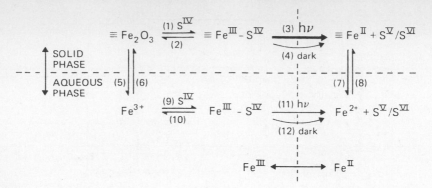

Figure 5. Conceptual model of iron redox chemistry in hematite suspensions containing S(IV) (Reproduced from Faust and Hoffmann, 1986).

release of soluble iron in the presence of oxalic acid (an important component of soils and sediments) followed by a much faster reaction. These authors propose a two-stage reaction: (a) comparatively slow release of Fe(III) through complexation by adsorbed oxalate, and (b) a faster, secondary, reductive dissolution step involving electron transfer from readsorbed Fe(II) (present at the oxide surface as ferrous oxalate) which is formed principally as a result of solution phase photoreduction of ferric oxalate. A schematic outline of the proposed mechanism of the fast reductive dissolution step is shown in Figure 6.

Although the role of oxidizable sorbents in the primary light absorption process appears to depend very much on the particular situation, the nature of these agents and the quantity sorbed at the particle surface will clearly be critical in determining the rate and extent of photoreduction of solid matrix metal ion. Once reduced, the tendency for metal species to be released to solution will depend, in part, on the nature of the oxide. (Leland and Bard, 1987; Waite and Torikov, 1987). For example, lepidocrocite would be expected (and is typically observed) to be more reactive than goethite because, unlike goethite, it has a relatively open layer structure consisting of sheets of Fe(O,OH)$_6$ octahedra held together by hydrogen bonds resulting in a greater proportion of metal atoms in active sites, i.e. sites having fewer structural bonds at the edges of the sheets

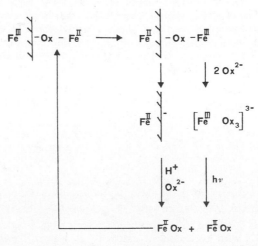

Figure 6. Schematic outline of the proposed mechanism of the fast step in the photochemical dissolution of goethite by oxalate (Reproduced from Cornell and Schindler, 1987).

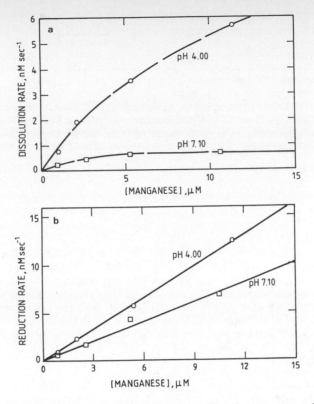

Figure 7.   (a) Dependence of observed initial dissolution rate of illuminated suspension of $MnO_2$ in 10 mg/L fulvic acid on manganese dioxide concentration at pH 4.0 and 7.1, and (b) Initial rate of Mn(IV) reduction as a function of $MnO_2$ concentration estimated from the results shown in (a) (Reproduced from Waite et al., 1988).

(Cornell and Schindler, 1987). The suspension pH will also exert a strong influence on the release, or otherwise, of reduced metal ions from the solid with an increasing tendency (in the absence of strong reduced metal ion ligands in solution) for ferrous and manganous ions to be retained at the respective oxide surfaces as the pH increases. The effect on dissolution rate of the increasing affinity of cationic species for increasingly negative surfaces as suspension pH is raised is clearly shown in the results of studies into the photoassisted dissolution of a colloidal manganese oxide in the presence of fulvic acid (Figure 7).

**Abiotic Nitrogen Reduction**

As indicated above, photolysis of semiconducting metal oxides with light of energy greater than the oxide band gap should result in hole-induced oxidation of adsorbed species with concomitant reduction reactions. Thus, a number of reports of $N_2$ reduction to $NH_3$ when $TiO_2$ based catalysts dispersed in water are irradiated with UV light are documented in the literature (Schrauzer and Guth, 1977; Miyama et al., 1980; Schrauzer et al., 1983; Grayer and Halmann, 1984). Of far greater significance to natural water chemistry is the recent report by Tennakone et al. (1987) that photolysis of an aqueous suspension of hydrous ferric oxide $[Fe_2O_3(H_2O)_n]$ with visible light resulted in photo-reduction of $N_2$ to $NH_3$ with concomitant oxidation of water. Tennakone et al. (1987) consider that the catalytic activity of the amorphous iron oxyhydroxide originates from its strongly negative flat band potential and chemisorption of $N_2$ with associated weakening of the N-N bond.

The reactions of the photogenerated electrons and holes are summarized in equations (1) and (2).

$$N_2 + 6H^+ + 6e^- \longrightarrow 2NH_3 \tag{1}$$

$$3H_2O + 6h^+ \longrightarrow {}^3/_2O_2 + 6H^+ \tag{2}$$

Tennakone et al. (1987) found the molar ratio of the yield of $NH_3$ to $O_2$ from samples irradiated for a short period of time (10-20 minutes) to be in approximate agreement with the stoichiometric value ($NH_3:O_2=1:0.75$) expected from the above equations. Kinetically, reaction (1) is favoured by a low pH, however, the highly negative flat band

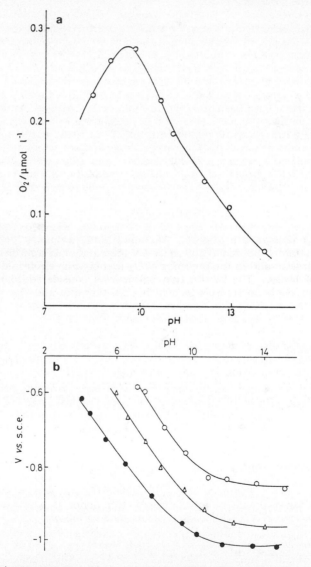

Figure 8.    (a) pH variation of the initial reaction rate in the iron oxyhydroxide photocatalyzed reduction of $N_2$ and concomitant oxidation of $H_2O$, and (b) pH variation of the photocurrent onset potential [in V vs. saturated calomel electrode (s.c.e.)] for $Fe_2O_3(H_2O)_n$ (●), α-$Fe_2O_3$ (o), and $Fe_2O_3(H_2O)_n$ (Δ) in the presence of $NO_3^-$ (ca. 10 μmol/L) (Reproduced from Tennakone et al., 1987).

potential needed for nitrogen reduction is realized only under alkaline conditions. The variation of the reaction rate with pH, i.e. existence of an optimal pH as shown in Figure 8a, appeared to be caused by the above effect.

Interestingly, these authors found that other iron oxides such as hematite do not possess the ability to photoreduce $N_2$ or $H_2O$, presumably because of less negative flat band potentials (see Figure 8b). The ability of the amorphous iron oxide to catalyse the reduction of nitrogen was also found to be lower in the presence of trace quantities of nitrate which shifted the flat band potential for this oxide in the positive direction. The catalytic activity of the amorphous iron oxide seemed to depend on the high degree of hydration as any attempt to dry the catalyst (even freeze drying) denatured it completely.

## Hydrogen Peroxide Production

Hydrogen peroxide, $H_2O_2$, is one of the most powerful oxidants now recognized to be present in natural waters. Concentrations as high as 60 µM for rainwater and 250 µM for cloudwater samples have been reported (Richards et al., 1983; Kelly et al., 1985; Zika et al., 1982). Peroxide concentrations in the range 10-300 nM are more typical for surface marine waters (Zika et al., 1985) and low micromolar concentrations of $H_2O_2$ have been reported for freshwaters (Cooper and Zika, 1983; Draper and Crosby, 1983). Although some blue-green algae have been observed to produce hydrogen peroxide (Stevens et al., 1973), the major pathway for production of $H_2O_2$ in natural waters appears to involve the photooxidation of natural organic matter with initial formation of the superoxide radical anion (or its conjugate acid, the hydroperoxyl radical) and subsequent disproportionation of this intermediate to hydrogen peroxide (Petasne and Zika, 1987; Zepp et al., 1987). The scavenging by water droplets of $HO_2 \cdot$ radicals produced in the gas phase also appears to be an important mode of $H_2O_2$ generation in cloudwater droplets (Graedel and Goldberg, 1983).

In addition to the pathways for hydrogen peroxide formation mentioned above, several other mechanisms are possible. Hoffmann and coworkers (Bahnemann et al., 1987; Kormann et al., 1988) report that near-UV illumination of aqueous suspensions of ZnO, $TiO_2$ and desert sand in the presence of $O_2$ and organic electron donors results in the formation of $H_2O_2$. This production pathway is considered to be of particular significance in atmospheric aerosols for which iron, titanium and zinc are amongst the most abundant metals detected in ambient samples and for which organically mediated pathways are likely to be less important (Bahnemann et al., 1987).

A semiconducting mechanism identical to that described above has been proposed to account for the observed production of $H_2O_2$. As previously described, photolysis of the semiconducting metal oxide with light of energy above the band gap energy generally leads to the formation of an electron-hole pair in the semiconducting particle. $H_2O_2$ can be formed via two different pathways in an aerated aqueous solution provided that $e^-$ and $h^+$ are generated:

$$O_2 + 2e^- + 2H^+ \longrightarrow H_2O_2 \qquad (3)$$

$$2H_2O + 2h^+ \longrightarrow H_2O_2 + 2H^+ \qquad (4)$$

Appreciable yields of hydrogen peroxide are detected only when appropriate electron donors, D, are added before illumination. The electron donor, D, which must be adsorbed on the particle surface, reacts with a valence band hole as follows:

$$D + h^+ \longrightarrow D^+ \qquad (5)$$

Electron donors bound to the surface of the semiconductor particles interfere with $e^-$-$h^+$ recombination allowing conduction band electrons to react with molecular oxygen via reaction (3).

Kormann et al. (1988) obtained steady-state concentrations of $H_2O_2$ in excess of 100 µM on illumination of suspensions of ZnO and report quantum yields for peroxide formation of *ca.* 15% at 330 nm. On the other hand, only low hydrogen peroxide concentrations were found in illuminated $TiO_2$ suspensions, presumably because of the high stability of Ti-peroxo complexes. Hydrogen peroxide formed on the $TiO_2$ surface is not readily released and is thus more susceptible to degradation processes at the particle surface. Desert sand suspensions exhibited photochemical behaviour similar to that of $TiO_2$ and ZnO although thermal degradation of $H_2O_2$ appeared to be relatively rapid, possibly as a result of a "Fenton-type" reaction between $H_2O_2$ and Fe(II) present in the desert sand.

## HETEROGENEOUS PHOTOCHEMICAL TREATMENT OF WATERS AND WASTES

### Oxidative Degradation of Toxic Substances

**Kinetic and Mechanistic Aspects of Degradation Processes.** One of the most potentially useful applications of semiconducting particles is in the well-documented photocatalyzed degradation of toxic substances in waters and wastes (Pelizzetti and Serpone, 1986). Essential aspects of this process are aptly demonstrated by the results of studies by Ollis and coworkers on the degradation of a series of chloro and bromo hydrocarbons, including the ubiquitous, carcinogenic water contaminants trichloromethane (chloroform) and trichloroethylene (TCE), in the presence of a $TiO_2$ catalyst (Pruden and Ollis, 1983; Hsiao et al., 1983; Ahmed and Ollis, 1984; Ollis, 1985). As shown in Figure 9, degradation is only observed in the presence of both illumination plus catalyst. In the case of the chlorohydrocarbons, complete mineralization to $CO_2$ and HCl was observed, with stoichiometric formation of chloride ion. For the photocatalyzed degradation of TCE, the overall stoichiometry for the reaction can be written as follows:

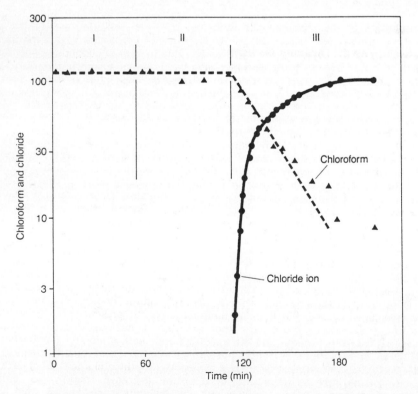

Figure 9.    Photocatalyzed degradation of chloroform (initial chloroform conc. = 122 ppm, initial chloride = 2 ppm). Region I - illumination only, Region II - catalyst only, Region III - illumination plus catalyst (Reproduced from Ollis, 1985).

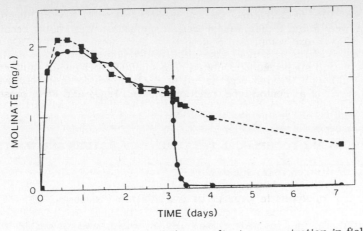

Figure 10. Effect of ZnO (14 kg Zn/ha) on molinate concentration in field water (●) when applied on Day 3 (↓) compared to an untreated control (■) (Reproduced from Draper and Crosby, 1987).

$$CCl_2CHCl + H_2O + 3/2O_2 \longrightarrow 2CO_2 + 3HCl \qquad (6)$$

For all halogenated hydrocarbons examined, Ollis and coworkers found the initial rate of reaction *vs.* reactant concentration to be satisfactorily modelled by a Langmuirian sorption process followed by a slow surface reaction step. In this case, the rate of reaction is proportional to the coverage of the surface, $\theta_x$, by an adsorbed intermediate, x:

$$rate = k.\theta_x \qquad (7)$$

where $\theta_x$ is given by the Langmuir isotherm:

$$\theta_x = \frac{K.c}{1 + K.c} \qquad (8)$$

where c is the solution reactant concentration and K is the apparent binding constant of the intermediate on the illuminated catalyst.

In addition to the photodegradation of halogenated aliphatics, many other toxic or otherwise problematic organic compounds for which biodegradation and bioelimination through bacterial processes is slow may be rapidly photodegraded in the presence of semiconductor-based catalysts. Details of the photocatalyzed degradation of a wide range of aromatic compounds have been reported including phenol (Okamato et al., 1985), polychlorinated biphenyl (Carey et al., 1976), chlorobenzene (Oliver et al., 1979), and chlorophenols (Barbeni et al., 1984; Matthews, 1986 and 1987; Bonhomme and Lemaire, 1986; and Pichat, 1987). These simple compounds are typically mineralized completely via a range of intermediate products. For example, the intermediates hydroquinone, pyrocatechol, 1,2,4-benzenetriol, pyrogallol, and 2-hydroxy-1,4-benzoquinone are initially formed in the initial stages of photolysis of phenol but are further oxidized via acids and/or aldehydes to $CO_2$ and $H_2O$. The surfactant sodium dodecylbenzenesulphonate (DBS) has also been shown to degrade on illumination in the presence of titanium dioxide particles with the aromatic group (and the surfactant behaviour) decaying within one hour under exposure to sunlight and the long aliphatic chain undergoing oxidation more slowly (Hidaka et al., 1986). Draper and Crosby (1987) report that the widely used herbicides molinate (S-ethyl hexahydro-1H-azepine-1-carbothioate) and thiobencarb (S[(4-chlorophenyl)]methyl-N,N-diethylcarbamothioate) are degraded rapidly in sunlight-irradiated suspensions of $TiO_2$ and ZnO, with ZnO exhibiting particularly high activity (Figure 10). A range of relatively stable oxidation products were detected by gas chromatography/mass spectrometry, but none persisted and none are presently recognized as long term hazards.

Although reaction mechanisms are case specific, the photocatalyzed mineralization of sorbed species appears to occur generally via the hydroxyl radical, ·OH. For example, the degradation of TCE, which occurs via the intermediate dichloroacetaldehyde ($Cl_2HCCHO$), is satisfactorily described by the reaction scheme shown in Figure 11. The dichloroacetaldehyde (DCA) is most likely decomposed by further reaction with hydroxyl radicals (Pruden and Ollis, 1983). In many cases it is likely that $H_2O_2$ will also be formed via reduction of sorbed oxygen by photogenerated electrons. There are suggestions that the resultant peroxide could undergo further reduction to ·OH which may then participate in the oxidation process (Okamoto et al., 1986).

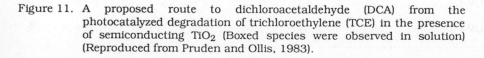

Figure 11. A proposed route to dichloroacetaldehyde (DCA) from the photocatalyzed degradation of trichloroethylene (TCE) in the presence of semiconducting $TiO_2$ (Boxed species were observed in solution) (Reproduced from Pruden and Ollis, 1983).

In addition to the possibilities for degradation of organic pollutants in waters and wastes, some consideration has been given to the use of photolyzed semiconductor suspensions in the oxidation of inorganic contaminants. Particular attention has been given to the oxidation of cyanide, CN⁻, to the less toxic forms of cyanate, OCN⁻, and thiocyanate, SCN⁻ (Serpone et al., 1988). Thus, Frank and Bard (1977a and b) demonstrated the transformation of CN⁻ to OCN⁻ in anatase suspensions using xenon arc lamp irradiation (obtaining quantum efficiencies *ca.* 6%) while Borgarello et al. (1986) obtained high quantum efficiencies (>25%) for the conversion of CN⁻ to SCN⁻ on photolysis of Rh-loaded CdS dispersions with visible light. It has been suggested that these photocatalytic methods have particular potential in the treatment of wastes containing high concentrations of cyanide such as those generated in coal gasification, electroplating and gold processing (Serpone et al., 1988).

**Economic Aspects**. It is clear from the above discussion and the numerous examples presented, that heterogeneous photocatalysis represents a powerful method for the degradation of a wide range of toxic or otherwise troublesome water contaminants.

However, to assess the commercial potential of this apparently general process for water decontamination and water purification more fully, further studies of its ability to degrade complex, multicomponent wastes are needed. Considerable attention also needs to be devoted to reactor and process design with factors such as reactor type (e.g. batch *vs.* continuous) and illumination configuration clearly being critical. In addition, economic analyses are needed in which established technologies for contaminant removal are compared with heterogeneous photocatalysis and other emerging technologies of potentially similar advantage. An analysis of this type has been presented by Ollis (1987) in a comparison of processes suitable for the removal of a nonvolatile hazardous organic contaminant from waters; aspects of this analysis are presented below.

Ollis (1988) notes that among existing water purification processes, the United States Environmental Protection Agency regards only air stripping (for removal of volatile contaminants) and carbon adsorption (for removal of volatile and nonvolatile contaminants) as the "most effective treatment technologies", a category indicating both a high level of water purification and wide generality. However, a key weakness of these techniques is that they are nondestructive technologies. Thus, air strippers convert a liquid contamination problem into an air pollution problem, and carbon adsorption produces a hazardous solid which, at least in the United States after 1988, cannot be landfilled. Consequently, pressure for elimination of the pollution hazard will lead continually away from nondestructive processes toward destructive ones which should, ideally, produce a totally harmless effluent. Strategically then, heterogeneous photocatalysis will have as competition for new process development those processes which utilize a chemically destructive approach. The major processes of interest in this category, in addition to heterogeneous photocatalysis, are those of photolysis (UV illumination), ozone, UV-ozone and chlorine. Of these, only UV-ozone and photocatalytic processes can result in relatively rapid, complete destruction of most nonvolatile contaminants (Ollis, 1988).

Table 1. Operating Costs ($/thousand gallons) of Plants of Increasing Size (Megagallons/Day) for Treatment of a Waste Containing 50 ppb Polychlorinated Biphenyl (effluent < 1ppb ) Using Either Carbon Adsorption, UV Ozonation, or Heterogeneous Photocatalysis (Reproduced from Ollis, 1988)

| Plant Size | 0.029 | 0.058 | 0.115 | 0.23 | 0.46 | 0.92 | 2.44 |
|---|---|---|---|---|---|---|---|
| Carbon | 4.40 | 3.05 | 2.40 | 1.80 | 1.50 | 1.25 | 1.10 |
| UV-Ozone | 7.34 | 4.77 | 3.57 | 2.78 | 2.39 | 2.16 | 1.78 |
| UV-Photocatal. | 5.56 | 3.57 | 2.46 | 1.81 | 1.43 | 1.31 | 1.13 |

Details of the analysis undertaken by Ollis (1988) cannot be presented here, but the final summary of operating costs obtained for treatment of a waste containing a polychlorinated biphenyl (PCB) (feed = 50 ppb, effluent < 1 ppb) using carbon adsorption, UV-ozonation and heterogeneous photocatalysis are shown in Table 1 and indicate that, over most process scales considered, the photocatalytic process is competitive with carbon. While it must be stressed that many simplifying assumptions have been made in this analysis by Ollis, the results are encouraging and certainly validate the continuation of research in this area.

## Heavy Metals Precipitation

The light-assisted precipitation of metals onto semiconducting particles has recently been reported and the possibility of using semiconductor-catalyzed processes in the removal of toxic metals from waste streams has been suggested, as has the possibility of using these processes in the selective concentration and separation of precious metals (Serpone et al., 1987). For example, Tanaka et al. (1986) report that on illumination of a Pt-loaded $TiO_2$ suspension containing one millimolar $Pb^{2+}$, the $Pb^{2+}$ concentration in solution decreased by 86% in 15 minutes (no significant loss of $Pb^{2+}$ occurred on this time-scale in the dark or in the absence of $TiO_2$). The proportion of $Pb^{2+}$ removed was found to decrease on deoxygenation of the suspension. Based on these observations, Tanaka et al. (1986) suggest that the metal ions are oxidized by holes generated on illumination with a possible oxidizing contribution from superoxide arising from the reduction of sorbed oxygen by trapped electrons at the particle surface;

i.e. $$2O_2 + 2e^- \longrightarrow 2O_2^{2-} \tag{9}$$

$$Pb^{2+} + 2H_2O + 2h^+ \longrightarrow PbO_2 + 4H^+ \tag{10}$$

$$Pb^{2+} + 2O_2^- \longrightarrow PbO_2 + O_2 \tag{11}$$

or

$$Pb^{2+} + 2O_2^- \longrightarrow PbO + {}^3/_2O_2 \tag{12}$$

When metal-free $TiO_2$ was used in place of Pt-loaded $TiO_2$, PbO was deposited instead of $PbO_2$, though the efficiency of the deposition was much smaller than in the presence of Pt-loaded $TiO_2$. It thus appears that the interaction of $Pb^{2+}$ with $O_2^-$ radical occurs predominantly by reaction (12).

Tanaka et al. (1986) also found that $Hg^{2+}$ was efficiently eliminated from a $10^{-4}$ M $Hg^{2+}$ solution with 76% and 83% deposited onto $TiO_2$ and $TiO_2$/Pt respectively. The proportion of $Hg^{2+}$ deposited in each case increased slightly on deoxygenation. From these results and its redox potential ($E^o_{(Hg^{2+}/Hg^o)}$ = 0.851 V $vs$. NHE), it is reasonable to consider that $Hg^{2+}$ is reduced by the electron from the conduction band and water is oxidized by the positive hole. This consideration is supported by the observation that oxygen evolved as $Hg^{2+}$ was deposited (Tanaka et al., 1986). A similar mechanism has been proposed by Domenech and Prieto (1986) to account for the precipitation of $Cu^o$ on ZnO particles on photolysis of a suspension containing $Cu^{2+}$. These workers consider that this photocatalyzed reduction process exhibits real potential as a future waste treatment method.

## CONCLUSIONS

There is an extensive literature on photoprocesses in natural waters and application of photoprocesses to the treatment of waters and wastes. Recent studies have yielded considerable insight into the range of light-driven processes that occur in the natural aquatic environment or that result in degradation of toxic or troublesome substances in waters and wastes. In addition, detailed laboratory studies in simple, well-defined systems have provided information on the likely kinetics and mechanism of photoprocesses in complex natural waters and wastes. Particular attention has been devoted in this chapter to heterogeneous photoprocesses which are observed in, or may have significance for, the natural aquatic environment.

The photoinduced or photoassisted dissolution of iron and manganese oxides in natural waters is now well documented and may have considerable implications for certain systems, either because of the alteration in the ability of these oxides to scavenge toxic (e.g. trace metals or organic compounds) or otherwise troublesome (e.g. phosphorus) species or because of the increased supply of the potentially limiting nutrients iron and manganese to the biota. The photodissolution of metal oxides in natural waters is a multistep process involving a sequence of surface reactions including, typically, adsorption of naturally occurring organic compounds, electron transfer processes at the particle surface and release of oxidised and reduced species to solution. The nature of the primary chromophore is unclear at this stage but may involve absorption by either the solid oxide (a semiconductor mechanism), the adsorbed surface species or both. Abiotic reduction of nitrogen and formation of hydrogen peroxide have also been observed on photolysis of aqueous suspensions of semiconducting metal oxides and there have been suggestions that these processes may be important in certain aquatic environments.

Widespread interest has been shown in the possibility of using semiconducting particles in the degradation of toxic or troublesome organic compounds such as herbicides, pesticides and surfactants. Heterogeneous photocatalytic degradation of wastes is particularly attractive because of the efficiency with which it can degrade organic species to harmless inorganic components (i.e. complete mineralization can be obtained) using relatively inexpensive, easily handled reagents such as titanium dioxide. Economic analysis of the heterogeneous photocatalysis option indicates that treatment costs are likely to be similar to the costs associated with carbon adsorption though considerably more information on the effectiveness of this process in treating complex, multicomponent wastes is needed. The possibility of using semiconducting particles in the photoinduced removal of trace elements such as lead, copper and mercury from wastes has also been raised.

## REFERENCES

Ahmed, S. and Ollis, D.F., 1984. Solar photoassisted catalytic decomposition of the chlorinated hydrocarbons trichloroethylene and trichloromethane, *Solar Energy*, 32:597.

Anderson, M.A. and Morel, F.M.M., 1983. The influence of aqueous iron chemistry on the uptake of iron by the coastal diatom *Thalassiosira weissflogii*, *Limnol. Oceanogr.*, 27:789.

Bahnemann, D.W., Hoffmann, M.R., Hong, A.P. and Kormann, C., 1987. Photocatalytic formation of hydrogen peroxide, in: "The Chemistry of Acid Rain: Sources and Atmospheric Processes," ACS Symposium Series No. 349, American Chemical Society, Washington, DC., p 120.

Barbeni, M., Pramauro, E., Pelizzetti, E., Borgarello, E., Gratzel, M. and Serpone, N., 1984. Photodegradation of 4-chlorophenol catalyzed by titanium dioxide particles, *Nouv. J. deChim.*, 8:547.

Bonhomme, G. and Lemaire, J., 1986. Hydroxylation du chloro-3-phenol photocatalysee par l'oxyde de zinc, *C.R. Acad. Sci. Paris, II*, 302:769.

Borgarello, E., Terzian, R., Serpone, N., Pelizzetti, E. and Barbeni, M., 1986. Photocatalyzed transformation of cyanide to thiocyanate by rhodium-loaded cadmium sulphide in alkaline aqueous sulphide media, *Inorg. Chem.*, 25:2135.

Brand, L.E., Sunda, W.G. and Guillard, R.R.L., 1983. Limitation of marine phytoplankton reproduction rates by zinc, manganese and iron, *Limnol. Oceanogr.*, 28:1182.

Carey, J.H., Lawrence, J. and Tosine, H.M., 1976. Photodechlorination of PCB's in the presence of titanium dioxide in aqueous suspensions, *Bull. environ. contam. Toxicol.*, 16:697.

Collienne, R.H., 1983. Photoreduction of iron in the epilimnion of acidic lakes, *Limnol. Oceanogr.*, 28:83.

Cooper, W. and Zika, R.G., 1983. Photochemical formation of hydrogen peroxide in surface and groundwaters exposed to sunlight, *Science*, 220:711.

Cornell, R.M. and Schindler, P.W., 1987. Photochemical dissolution of goethite in acid/oxalate solution, *Clays & Clay Minerals*, 35:347.

Domenech, J. and Prieto, A., 1986. Photoelectrochemical reduction of Cu(II) ions in illuminated aqueous suspensions of ZnO, *Electrochim. Acta*, 31:1317.

Draper, W.M. and Crosby, D.G., 1983. The photochemical generation of hydrogen peroxide in natural waters, *Arch. Environ. Contam. Toxicol.*, 12:121.

Draper, R.B. and Crosby, D.G., 1987. Catalyzed photodegradation of the herbicides molinate and thiobencarb, *in:* "Photochemistry of Environmental Aquatic Systems," R.G. Zika and W.J. Cooper, eds., ACS Symposium Series No. 327, American Chemical Society, Washington, DC., p 240.

Entsch, B., Sim, R.G. and Hatcher, B.G., 1983. Indications from photosynthetic components that iron is a limiting nutrient in primary producers on coral reefs, *Mar. Biol.*, 73:17.

Faust, B.C. and Hoffmann, M.R., 1986. Photoinduced dissolution of $\alpha$-$Fe_2O_3$ by bisulphite, *Environ. Sci. Technol.*, 20:943.

Finden, D.A.S., Tipping, E., Jaworski, G.H.M. and Reynolds, C.S., 1984. Light-induced reduction of natural iron(III) oxide and its relevance to phytoplankton, *Nature*, 309:783.

Frank, S.N. and Bard, A.J., 1977a. Heterogeneous photocatalytic oxidation of cyanide and sulphite in aqueous solutions at semiconductor powders, *J. Phys. Chem.*, 81:1484.

Frank, S.N. and Bard, A.J., 1977b. Heterogeneous photocatalytic oxidation of cyanide ion in aqueous solutions at $TiO_2$ powder, *J. Amer. Chem. Soc.*, 99:303.

Frese, K.W. and Canfield, D., 1983. Reduction of $CO_2$ by n-type semiconductor electrons, *in:* "Extended Abstracts, 153rd Meeting of the Electrochemical Society, Seattle, WA", Electrochemical Society, Pennington, NJ, Abstract No. 517, p 784.

Graedel, T.E. and Goldberg, K.I., 1983. Kinetic studies of raindrop chemistry, *J. Geophys. Res.*, 88:865.

Grayer, S. and Halmann, M., 1984. Electrochemical and photoelectrochemical reduction of molecular nitrogen to ammonia, *J. Electroanal. Chem.*, 170:363.

Hidaka, H. Kubota, H., Gratzel, M., Pelizzetti, E., and Serpone, N., 1986. Photodegradation of surfactants II: Degradation of sodium dodecylbenzene sulphonate catalyzed by titanium dioxide particles, *J. Photochem.*, 35: 219.

Hsiao, C-Y., Lee, C-L. and Ollis, D.F., Heterogeneous photocatalysis: Degradation of dilute solutions of dichloromethane ($CH_2Cl_2$), chloroform ($CHCl_3$), and carbon tetrachloride ($CCl_4$) with illuminated $TiO_2$ photocatalyst, *J. Catal.*, 82:418.

Kelly, T.J., Daum, P.H. and Schwartz, S.E., 1985. Measurements of peroxides in cloudwater and rain, *J. Geophys. Res.*, 90:7861.

Kormann, C., Bahnemann, D.W. and Hoffmann, M.R., 1988. The photocatalytic production of $H_2O_2$ and organic peroxides in aqueous suspensions of $TiO_2$, ZnO, and desert sand, *Environ. Sci. Technol.*, 22:798.

Langford, C.H. and Carey, J.H., 1987. Photocatalysis by inorganic components of natural water systems, *in:* "Photochemistry of Environmental Aquatic Systems," R.G. Zika, and W.J. Cooper, eds, ACS Symposium Series No. 327, American Chemical Society, Washington, DC., p 225.

Leland, J.K. and Bard, A.J., 1987. Photochemistry of colloidal semiconducting iron oxide polymorphs, *J. Amer. Chem. Soc.*, 91:5076.

Litter, M.I. and Blesa, M.A., 1988. Photodissolution of iron oxides: I. maghemite in EDTA solutions, *J. Coll. Interface Sci.*, 125:679.

Martin, J.H. and Gordon, R.M., 1988. Northeast Pacific iron distributions in relation to phytoplankton productivity, *Deep-Sea Research*, 35:177.

Martin, J.H. and Fitzwater, S.E., 1988. Iron deficiency limits phytoplankton growth in the northeast Pacific subarctic, *Nature*, 331:341.

Matthews, R.W., 1986. Photo-oxidation of organic material in aqueous suspensions of titanium dioxide, *Water Res.*, 20:569.

Matthews, R.W., 1987. Carbon dioxide formation from organic solutes in aqueous suspensions of ultraviolet-irradiated $TiO_2$. Effect of solute concentration, *Aust. J. Chem.*, 40:667.

McKnight, D.M., Kimball, B.A. and Bencala, K.E., 1988. Iron photoreduction and oxidation in an acidic mountain stream, *Science*, 240:637.

Miyama, H., Fujii, N. and Nagal, Y., 1980. Heterogeneous photocatalytic synthesis of ammonia from water and nitrogen, *Chem. Phys. Lett.*, 74:523.

Morel, F.M.M., 1983. "Principles of Aquatic Chemistry," Wiley, New York.

Okamaoto, K., Yamamoto, Y., Tanaka, H., Tanaka, M. and Itaya, A., 1985. Heterogeneous photocatalytic decomposition of phenol over $TiO_2$ powder, *Bull. Chem. Soc. Jap.*, 58:2015.

Oliver, B.G., Cosgrove, E.G. and Carey, J.H., 1979. Effect of suspended sediments on the photolysis of organics in water, *Environ. Sci. Technol.*, 13:1075.

Ollis, D.F., 1985. Contaminant degradation in water, *Environ. Sci. Technol.*, 19:480.

Ollis, D.F., 1987. Process economics for water purification: A comparative assessment, *in:* "Proceedings Conference on Photocatalysis and Photoelectrochemistry", *NATO ASI Ser. Ser. C,* 237:663.

Pelizzetti, E. and Serpone, N., 1986. "Homogeneous and Heterogeneous Photocatalysis," D.Reidel, Dordrecht, Holland.

Petasne, R.G. and Zika, R.G., 1987. Fate of superoxide in coastal seawater, *Nature,* 325:516.

Pichat, P., 1987. Powder photocatalysts: characterization by isotopic exchanges and photoconductivity; potentialities for metal recovery, catalyst preparation and water pollutant removal, *in:* "Proceedings Conference on Photocatalysis and Photoelectrochemistry", *NATO ASI Ser. Ser. C,* 237:399.

Pruden, A.L. and Ollis, D.F., 1983. Photoassisted heterogeneous catalysis: The degradation of trichloroethylene in water, *J. Catal.,* 82:404.

Richards, L.W., Anderson, J.A., Blumenthal, D.L., McDonald, J.A. and Kok, G.L., 1983. Hydrogen peroxide and sulphur(IV) in Los Angeles cloud water, *Atmos. Environ.,* 17:911.

Schrauzer, G.N. and Guth, T.D., 1977. Photolysis of water and photoreduction of nitrogen on titanium dioxide, *J. Amer. Chem. Soc.,* 99:7189.

Schrauzer, G.N., Strampach, N., Hui, L.N., Palmer, M.R. and Salehi, J., 1983. Nitrogen photoreduction on desert sands under sterile conditions, *Proc. Natl. Acad. Sci. USA,* 50: 3873.

Serpone, N., Borgarello, E. and Pelizzetti, E., 1987. Photoreduction and photodegradation of inorganic pollutants: II. Cyanides, *in:* "Proceedings Conference on Photocatalysis and Photoelectrochemistry", *NATO ASI Ser. Ser. C,* 237:499.

Sherman, D.M. and Waite, T.D., 1985. Electronic spectra of $Fe^{3+}$ oxides and oxide hydroxides in the near IR to near UV, *Amer. Mineral.,* 70:1262.

Stevens, S.E., Patterson, C.O.P. and Myers, J., 1973. The production of hydrogen peroxide by blue-green algae: a survey, *J. Phycol.,* 9:427.

Stramel, R.D. and Thomas, J.K., 1986. Photochemistry of iron oxide colloids, *J. Coll. Interface Sci.,* 110:121.

Stumm, W. and Morgan, J.J., 1981. "Aquatic Chemistry," 2nd ed., Wiley-Interscience, New York.

Sunda, W.G., Huntsman, S.A. and Harvey, G.R., 1983. Photoreduction of manganese oxides in seawater and its geochemical and biological implications, *Nature,* 301:234.

Sunda, W.G., 1989. Trace metal interactions with marine algae, *in:* "Marine Photosynthesis," R. Alberte, and R.T. Barber, eds., Oxford University Press, New York.

Tanaka, K., Harada, K. and Murata, S., 1986. Photocatalytic deposition of metal ions onto $TiO_2$ powder, *Solar Energy,* 36:159.

Tennakone, K., Wickramanayake, S., Fernando, C.A.N., Ileperuma, O.A. and Punchihewa, S., 1987. Photocatalytic nitrogen reduction using visible light, *J. Chem. Soc., Chem. Commun.,* 1987:1078.

Waite, T.D., 1986. Photoredox chemistry of colloidal metal oxides, *in:* "Geochemical Processes at Mineral Surfaces," J.A. Davis, and K.F. Hayes, eds, ACS Symposium Series No. 323, American Chemical Society, Washington, DC., p 426.

Waite, T.D. and Morel, F.M.M., 1984a. Photoreductive dissolution of colloidal iron oxides in natural waters, *Environ. Sci. Technol.,* 18:860.

Waite, T.D. and Morel, F.M.M., 1984b. Photoreductive dissolution of colloidal iron oxide: Effect of citrate, *J. Coll. Interface Sci.,* 102:121.

Waite, T.D., Wrigley, I.C. and Szymczak, R., 1988. Photo-assisted dissolution of a colloidal manganese oxide in the presence of fulvic acid, *Environ. Sci. Technol.,* 22:778.

Waite, T.D. and Torikov, A., 1987. Photoassisted dissolution of colloidal iron oxides by thiol-containing compounds, *J. Coll. Interface Sci.*, 119:228.

Zafiriou, O.C., 1983. Natural water photochemistry, *in:* "Chemical Oceanography," Vol. 8, Academic Press, London, pp 339-379.

Zafiriou, O.C., Joussot-Dubien, J., Zepp, R.G. and Zika, R.G., 1984. Photochemistry of natural waters, *Environ. Sci. Technol.*, 18:358A.

Zepp, R.G., 1980. Assessing the photochemistry of organic pollutants in aquatic environments, *in:* "Dynamics, Exposure and Hazard Assessment of Toxic Chemicals, "R. Haque, ed., Ann Arbor Science Publishers, Ann Arbor, Michigan, p 69.

Zepp, R.G., Braun, A.M., Hoigne, J. and Leenheer, J.A., 1987. Photoproduction of hydrated electrons from natural organic solutes in aquatic environments, *Environ. Sci. Technol.*, 21:485.

Zepp, R.G. and Wolfe, N.L., 1987. Abiotic transformation of organic chemicals at the particle-water interface, *in:* "Aquatic Surface Chemistry: Chemical Processes at the Particle-Water Interface," W. Stumm, ed., Wiley, New York, p 423.

Zika, R., 1981. Marine organic photochemistry, *in:* "Marine Organic Chemistry," E.K. Duursma, and R. Dawson, eds., Elsevier, Amsterdam, p 299.

Zika, R.G., Saltzman, E., Chameides, W.L. and Davis, D.D., 1982. $H_2O_2$ levels in rainwater collected in South Florida and the Bahama Islands, *J. Geophys. Res.*, 87:5015.

Zika, R.G., Moffett, J.W., Petasne, R.G., Cooper, W.J. and Saltzman, E.S., 1985. Spatial and temporal variations of hydrogen peroxide in Gulf of Mexico waters, *Geochim. Cosmochim. Acta*, 49:1173.

Zika, R.G. and Cooper, W.J., 1987. "Photochemistry of Environmental Systems," ACS Symposium Series No. 327, American Chemical Society, Washington, DC.

KINETICS AND MECHANISMS OF

IRON COLLOID AGGREGATION IN ESTUARIES

Keith A. Hunter

Chemistry Department
University of Otago
Dunedin, NEW ZEALAND

## INTRODUCTION

Estuaries play an important role in determining the flux of riverborne constituents reaching the ocean owing to their unique physico-chemical properties.  They are particularly important in effecting changes in the physical and chemical forms of different substances present in freshwater flows.  Many natural constituents of river waters are colloidal in nature, and therefore may undergo salt-induced aggregation processes within estuaries.  Two major components of riverine colloidal matter are organic humic and fulvic acids, and oxides of iron. This "brew" is (at least superficially) similar to proprietary flocculating mixtures used in water and waste stream treatment processes. Thus the study of natural aggregation in estuaries is complementary to an understanding of colloid behaviour in water treatment.

The iron-containing colloids in natural freshwaters are important support phases for many other dissolved chemical components.  In particular, they are able to sequester many surface-active organic pollutants such as chlorinated pesticides, solvents, polynuclear aromatic hydrocarbons, plasticizers and detergents.  An understanding of the geochemical behaviour of the colloids is therefore important in the assessment of waste water impacts on the aquatic environment. Aggregation of a colloid support phase in the saline zone of an estuary provides a mechanism for transfer of pollutants to sediments and an entry point to the food chain.

Iron-containing colloids in river and estuarine waters are conventionally separated from larger, filterable particles by membrane filtration, typically using 0.4-0.5 micron pore size filters.  This operational procedure, while somewhat arbitrary in nature, is widely used for water quality analysis.  The analytical amount of iron passing through such a filter is called "dissolved iron", and is known to exist mostly in the form of colloidal iron(III) oxides stabilized by association with humic substances (Shapiro, 1964; Sholkovitz, 1976; Moore et al., 1979).

Typical concentrations of dissolved iron colloids in river waters are in the range 1-10 µmol/L expressed as Fe.  By comparison, the concentrations of iron in seawater are extremely low.  Thus in an estuary, where the salinity increases from a few mg/L at the river end of the estuary to the seawater value of about 35 g/L, dissolved iron concentrations decrease steadily as river water is progressively diluted by seawater.

Figure 1 shows a typical profile of dissolved Fe as a function of salinity for the Taieri Estuary in New Zealand.  The marked downward curvature of the profile as salinity increases indicates that as seaward mixing progresses, the amount of dissolved iron is

*Surface and Colloid Chemistry in Natural Waters and Water Treatment*
Edited by R. Beckett, Plenum Press, New York, 1990

rapidly decreased. This behaviour is well documented and is an almost universal feature of estuarine mixing (Coonley et al., 1971; Windom et al., 1971; Bewers et al., 1974; Holliday and Liss, 1976; Boyle et al., 1977; Sholkovitz et al., 1978; Hunter, 1983).

During mixing of riverwater and seawater, riverine iron colloids are destabilized by increasing concentrations of seawater major ions and undergo aggregation through interparticle collisions (Sholkovitz, 1976; Eckert and Sholkovitz, 1976). This growth in particle size leads to a diminishing fraction of the colloidal size spectrum in the "dissolved" fraction size range.

## LABORATORY MIXING STUDIES

Laboratory studies in which river water and seawater are directly mixed in different proportions have been carried out. These generally show that aggregation takes place at variable rates depending on salinity and the time scale of observation (Mayer, 1982; Fox and Wofsy, 1983; Matsunaga et al., 1984, Hunter and Leonard, 1988). Typical results for such experiments are presented in Figure 2, where the fraction of dissolved iron in Taieri River water aggregated after increasing salinity to different values are plotted as a function of time.

These results show that a significant fraction of dissolved iron is aggregated more or less immediately after salt is added. This is followed by a steadily decreasing rate of aggregate formation over longer time scales, reaching a plateau after 40-100 min depending on salinity. Both the maximum amount of iron aggregated in the plateau region, and the amount aggregated initially, increase with salinity.

The variable rates of aggregation observed in the laboratory studies, and the dependence of these rates on salinity, complicate a kinetic analysis of the process

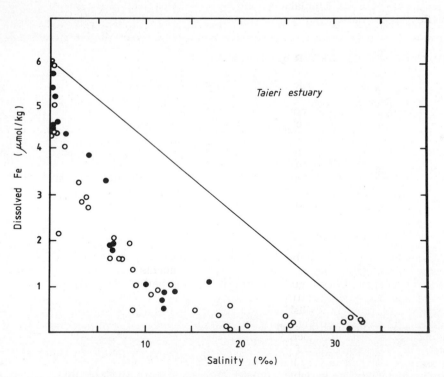

Figure 1.    Concentration of dissolved iron (Whatman GF/F filters) as a function of salinity for the Taieri Estuary, New Zealand (Data from Hunter (1983))

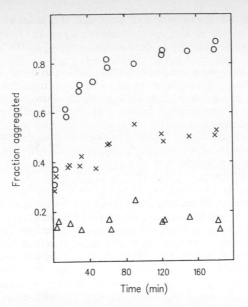

Figure 2.    Fraction of Taieri River dissolved iron aggregated as a function of time
after mixing with seawater.  Final salinities (g/L): 2.5 (triangles), 10.0
(crosses) and 15.0 (circles) (Data from Hunter and Leonard (1988))

through this kind of rate experiment.  Nonetheless, Mayer (1982) and Fox and Wofsy
(1983) have suggested that second-order kinetics can be used to describe at least part of
the process after the rapid initial step, and have used this to develop a quantitative model
of iron fluxes in estuaries.

## VALIDITY OF SECOND-ORDER KINETIC MODELS

The use of a second-order model is based principally on the belief that most of the
colloid aggregation takes place through Brownian collisions, a second-order process.
Strictly speaking, it is of more fundamental interest to focus on the kinetics of
interparticle collision in terms of particle number, rather than dissolved Fe concentration,
which is the sum of the mass of iron contained in all particles less than the pore size of
the filter.  Mayer (1982) has presented theoretical grounds for expecting second-order
particle number kinetics from Brownian aggregation to translate into second-order
removal of dissolved iron.  In more general terms, this correspondence is expected for any
self-preserving particle size spectrum (e.g. Hidy, 1965), but is probably an
oversimplification of the real situation (Farley and Morel, 1986).

Hunter and Leonard (1988) have suggested that the use of second-order and
related models has been somewhat overstated.  They point out that experimental artifacts
are present in much of the published work because the unstable colloids in river water
aggregate considerably before salt addition experiments are begun, changing the
subsequent kinetics.  They argue that in general, interparticle collision rates depend on
both the particle size spectrum and particle number concentrations, and are a
combination of different collision-producing mechanisms:  Brownian motion, shear
effects, and differential settling.  Most importantly, however, a kinetic analysis based on
likely collision models is rendered invalid if the efficiency of collisional encounter is not
constant, i.e. if the stabilities of the colloid particles are not uniform.  In a natural system
such as river water, it is very likely that the surface chemical and electrical properties
determining stability are variable.  Indeed, theoretical work suggests that stability also
depends on particle size (Overbeek, 1977) and is affected by variations in particle size and
surface charge in polydisperse, mixed suspensions (Prieve and Lin, 1982).

## EFFECTS OF ELECTROLYTE ON COLLOID STABILITY

Theories of colloid stability indicate that the effect of electrolyte on colloid aggregation may be divided into two concentration regimes (Overbeek, 1977). At low electrolyte concentration, interparticle electrostatic repulsion is promoted by the establishment of an electrical diffuse layer of ions surrounding each particle. As electrolyte concentration increases, this diffuse layer shrinks closer to the particle surface until the electrostatic barrier to close particle encounters disappears. When electrostatic repulsion is important, the efficiency of interparticle collisions in leading to aggregation is much less than 100% and is a sensitive function of electrolyte concentration (Ottewill and Shaw, 1965). In contrast, at high electrolyte concentration most collisions lead to aggregation and the aggregation kinetics become independent of electrolyte concentration (Overbeek, 1977).

For a purely electrostatically stabilized colloid, very little seawater is required to initiate aggregation with 100% collisional efficiency (at a salinity of 3 g/L, the Debye-Huckel double layer thickness is reduced to only about 1 nm). However the stability of iron colloids is dramatically increased through the steric stabilisation afforded by the presence of adsorbed organics (Napper, 1977), e.g. humic acids (Smith and Milne, 1981). It follows that the stability of riverine colloids is expected to be dependent on both electrolyte composition and the nature of the colloid surface, including adsorbed organics.

The appearance of a very rapid initial aggregation suggests that an important fraction of the dissolved iron is very unstable with regard to salt-induced aggregation. Hunter and Leonard (1988) have suggested that this is a consequence of complete, or nearly complete, electrolyte destabilisation of the colloids by the added seawater. They also suggest that the increasing fraction of iron removed as salinity increases suggests that there is a spectrum of inherent colloid stabilities in the particle population, with increasing numbers brought into the highly unstable regime as more salt is added. At the same time as this occurs, more stable colloids are aggregated at progressively lower rates over the observed time scales of several hours. Such a spectrum of particle stabilities may be regulated by particle size, shape or surface electrical properties (Prieve and Lin, 1982), or by differing degrees of stabilisation conferred by association with humic acids and specifically adsorbed ions (Hunter and Liss, 1979, 1982).

If this reasoning is correct, then a kinetic analysis of aggregation experiments in terms of simple collision models is ruled out by the changing stability ratios of the particles.

## THE IMPORTANCE OF LARGE PARTICLES

Large particles, similar to or larger in size that the filter used to separate the dissolved and filterable fractions, are known to be essential to the aggregation process. Figure 3 shows a comparison of the amount of "dissolved" iron aggregated from river water by addition of seawater with the same river water after all the filterable particles have been first removed. Prior removal of filterable particles causes a very large decrease in the rate and extent of iron aggregation, showing that the majority of collisions that remove colloidal iron-containing particles involve large particles. Hunter and Leonard (1988) have presented a simple statistical model of the aggregation process which accounts for this finding.

## DYNAMIC MIXING STUDIES

In the laboratory studies described so far, seawater is added directly to the riverine colloid suspension at the start of the experiment. This may bring some of the colloids into a salt-induced regime of instability that they would not experience in real estuaries. In real estuaries, river water and seawater seldom mix together directly as endmembers. Rather, mixing takes place in a continuous manner that may be approximated by a series of water types, each of which has a salinity close to that of its

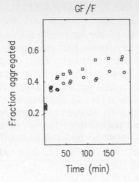

GF/F

Fraction aggregated

0.6

0.4

0.2

50  100  150

Time (min)

Figure 3.   Fraction of river water dissolved iron aggregated as a function of time after addition of seawater with and without prefiltration of river water. Open squares - no prefiltration; Open circles - prefiltered with Whatman GF/F. Final salinity 7.0 g/L, river water from Leith River, Dunedin (Data from Hunter and Leonard (1988))

neighbours.  If the mixing time is sufficiently long, many of the low-stability colloids will be removed from the spectrum at low salinities.  This prior removal would mean that there would be a different particle population undergoing aggregation at a given salinity from that being aggregated in the laboratory simulation where the salt addition is instantaneous.

The rapid initial aggregation suggests that colloid aggregation in estuaries is dependent on the history of freshwater/seawater mixing.  Accordingly, laboratory studies in which salt is gradually added to river water, as described originally by Smith and Longmore (1980), are a more realistic way of studying this process.  This type of dynamic mixing study has been considerably refined by Bale and Morris (1981).  These authors use a series of tanks interconnected with two-way pumps.  Seawater is pumped into the "rightmost" tank and river water into the "leftmost" tank.  The intermediate tanks then become filled with dynamic steady-state mixtures of river and seawater whose compositions depend on pumping rates and the composition of the neighbouring tanks.  The apparatus is esentially a physical version of a multi-box numerical model for an estuary.

Hunter et al. (1990) have used a much simpler experimental arrangement to investigate the kinetics of aggregation under dynamic conditions.  These authors used a single continuous-flow reactor containing fresh river water into which seawater was pumped at a known rate to gradually increase the salinity.  The overflow from the constant-volume reservoir was filtered and analysed for dissolved iron.

Typical iron-salinity profiles for Taieri River water and seawater mixed at flow rates corresponding to residence times in the reactor of between 490 and 65 min are presented in Figure 4.  The dissolved iron-salinity dependence is similar to that observed in field studies (e.g. Figure 1), showing that the same kind of aggregation process takes place in the laboratory reactor as in nature.  Interestingly, Carpenter and Smith (1986) have found the same results using a synthetic river water colloid prepared from amorphous iron oxide and humic acid.

## CATIONS RESPONSIBLE FOR AGGREGATION

Experiments using synthetic salt solutions in place of the seawater show that the aggregation process is initiated principally by divalent cations in the saline mixing endmember, principally calcium and magnesium ions.  While this is suggested by the classical Schultz-Hardy rule, calcium ions are about 5 times as effective as magnesium ions on a per mole basis, showing that the process giving rise to destabilisation of the colloids is not purely electrostatic in nature. Hunter et al. (1990) suggest that colloid destabilisation takes place principally through cation bridging of particles by ions such as calcium.

## KINETIC MODEL FOR DYNAMIC MIXING

Intuitively, one expects the kinetics of aggregation under the dynamic conditions of seawater-river water mixing in the reactor to be complex owing to the effects of changing salinity, dilution and other factors. It is therefore remarkable to find that the results can be described by a relatively simple kinetic law (Hunter et al., 1990).

At a constant flow rate v of seawater into the reactor vessel (volume V), the time rate of salinity increase is given by the first-order law

$$dS/dt = (v/V)(S_0 - S) \tag{1}$$

where S is salinity, $S_0$ is the seawater endmember salinity and the salinity of the river water has been approximated to zero for simplicity. The change in iron concentration due simply to dilution by the added seawater is

$$-d[Fe]/dt = k_0[Fe] \tag{2}$$

where $k_0 = (v/V)$ is the first-order rate constant for dilution which may be determined either from separate measurement of v and V or from a logarithmic plot of the freshwater fraction against time.

The overall decrease in dissolved iron concentration in the reactor with time is made up of two contributions - the decrease caused by dilution with seawater, and the decrease from aggregation effects, A.

$$-d[Fe]/dt = k_0[Fe] + A \tag{3}$$

The aggregation rate A may be a function of salinity, the rate of salinity increase, the rate of dilution, pH, and the concentrations of iron and natural organic matter such

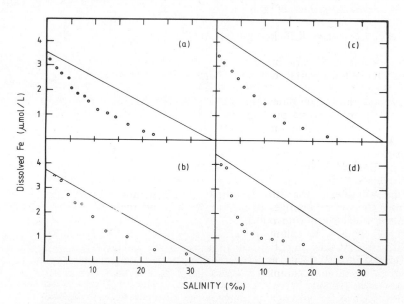

Figure 4.    Dissolved Fe-salinity profiles for model estuary experiments using Taieri River water and seawater endmembers at different pumping rates (Data from Hunter et al. (1990))

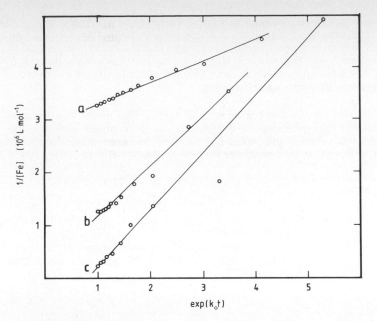

Figure 5.    Kinetic plots according to equation (4) of model estuary data obtained
using Leith River water  and seawater endmembers (Data from Hunter
et al. (1990))

as humic acid.  The problem of mathematically treating the kinetic results obtained with
the model estuary reactor is one of choosing an appropriate form for the aggregation term
A that allows integration of equation (3).

Hunter et al. (1990) argued that since the kinetics of aggregation in  simple mixing
studies were dominated, in rate terms, by the virtually instantaneous aggregation that
results from complete destabilisation of some of the colloids, the rate term A could be
approximated by a second-order expression independent of salinity.  In this case, setting
$A = k_2[Fe]^2$ allows integration of (3) to yield

$$1/[Fe] = 1/[Fe]_R(1+k_2[Fe]_R/k_0) \exp(k_0t) - (k_2/k_0) \tag{4}$$

where $k_2$ is a pseudo second-order rate constant for aggregation that is independent of
salinity and $[Fe]_R$ is the initial river water filterable iron concentration.

Figure 5 shows experimental data for a freshwater system, the Water of Leith, at a
number of different seawater pumping rates, plotted according to equation (4).  It is seen
that the data fit the equation quite well over the salinity range of the aggregation process.

**EFFECT OF SALINITY**

Figure 5 shows that the empirical rate constant for aggregation, $k_2$, is not
significantly affected by changes in salinity throughout most of the mixing range.  The
slopes of the kinetic plots do not significantly change with time whereas salinity does
(equation (1)).  Thus while salinity-dependent effects are observed in direct mixing of river
water and seawater, they do not appear to be important when the two solutions are mixed
together continuously.   This implies that the colloids aggregating at any particular
salinity must be already completely destabilized by the seawater electrolytes, i.e. fast
coagulation (Ives, 1978).

The main difference between the dynamic mixing experiments and those
performed by batch mixing is that in dynamic mixing, the dominant aggregation process

is fast coagulation, whereas in the batch experiments where salinity is fixed, only a fraction of the iron is aggregated in this way (Mayer, 1982; Hunter and Leonard, 1988) and the remainder undergoes aggregation at slower, salinity-dependent rates. The latter does not appear to be important in the model estuary because salinity is continuously increasing, bringing further colloids into the fast coagulation regime before a significant proportion has aggregated.

## EFFECT OF SEAWATER PUMPING RATE

The dynamic mixing studies show that aggregation increases linearly with the pumping rate into the reactor (Hunter et al, 1987) and can be represented empirically by

$$k_2 \cong k.k_0 \tag{5}$$

where the constant $k$ has a value of 0.5 L $\mu mol^{-1}$. This relationship is expected if the main aggregation process is fast coagulation.

## APPLICATION TO REAL ESTUARIES

A relationship between dissolved iron and salinity can be derived from the dynamic mixing equation (4) by using the relationship between salinity and time to substitute for t

$$1/[Fe] = 1/[Fe]_R(1 + k_2/k_0[Fe]_R) f^{-1} - k_2/k_0 \tag{6}$$

where $f = (S_0-S)/S_0$ is the freshwater fraction.

The form of equation (6) implies that the dissolved iron-salinity profile depends on the dimensionless parameter $\gamma = k_2/k_0[Fe]_R$. This suggests that in a simple estuary in which the rate of salinity increase is first-order, a steady-state dissolved iron-salinity profile will be developed which is independent of the rate of dilution. Figure 6 shows dissolved iron-salinity profiles calculated for various values of $\gamma$. The profiles have the same shape as actual field data, with $\gamma$ values in the range 2 to 16 covering the range in curvature exhibited by most estuaries.

Hunter et al. (1990) have demonstrated that equation (6) applies quite well to field data. This implicitly assumes that the rate of salinity increase is close to first-order. Figure 7 shows data plotted according to this equation for a number of estuaries in the USA, NZ and Australia. In spite of the assumptions involved, the agreement between model and field results is encouraging.

## CONCLUSIONS

Iron colloids in any river water aggregate at different rates when salt is added. In a continuous-flow reactor where salt addition takes place continuously over time scales of hours for doubling of salinity, fast aggregation dominates removal of the iron. This leads to the development of an iron-salinity profile which is independent of the doubling time for salinity. This result is capable of explaining the observation that iron-salinity profiles reported for most estuaries are broadly similar in spite of large differences in their time scales of salinity change. Furthermore, the mathematical description of kinetics in the model estuary seems to apply quite well to this type of field data, and gives derived kinetic data ($k_2/k_0$) of a similar magnitude to those measured in the laboratory.

This conceptual model for iron aggregation is probably applicable to most estuaries. However, it would not necessarily apply to well-mixed estuaries having very long salinity doubling times, because in this case salinity-dependent aggregation effects, resuspension and shear processes might be more important. Work is now progressing on applying the conceptual approach to an understanding of other colloids in river systems.

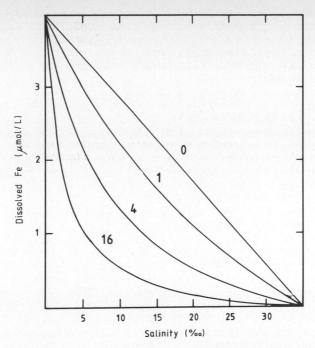

Figure 6.  Dissolved iron-salinity profiles calculated from equation (6). Numbers refer to the value of $\gamma = k[Fe]_R$ (Data from Hunter et al. (1990))

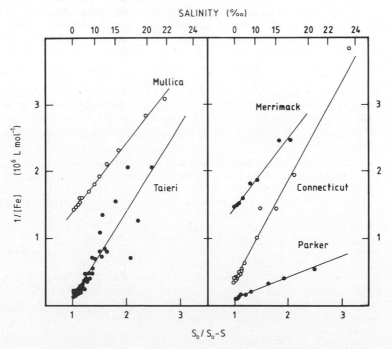

Figure 7.  Field results for estuarine dissolved iron profiles plotted according to equation [6] for 5 estuaries in USA and NZ (Data from Boyle et al. (1977) and Hunter (1983))

# REFERENCES

Bale, A.J. and Morris, A.W., 1981. Laboratory simulation of chemical processes induced by estuarine mixing: the behaviour of iron and phosphate in estuaries, *Estuar. Coast. Mar. Sci.*, 13:1-10.

Bewers, J.M., McCauley, I.D. and Sundby, B., 1974. Trace metals in the waters of the Gulf of St Lawrence, *Can. J. Earth Sci.*, 11:939-950.

Boyle, E.A., Edmond, J.M. and Sholkovitz, E.R., 1977. The mechanism of iron removal in estuaries, *Geochim. Cosmochim. Acta*, 41:1313-1324

Coonley, S., Baker, E.B. and Holland, H.D., 1971. Iron in the Mullica River and Great Bay, New Jersey, *Chem. Geol.*, 7a:51-63.

Eckert, J.M. and Sholkovitz, E.R., 1976. The flocculation of iron, aluminium and humates from river water by electrolytes, *Geochim. Cosmochim. Acta*, 40:847-848.

Farley, K.J. and Morel, F.M.M., 1986. Role of coagulation in the kinetics of sedimentation. *Environ. Sci. Technol.*, 20:187-195.

Fox, E.L. and Wofsy, S.C., 1983. Kinetics of removal of iron colloids from estuaries, *Geochim. Cosmochim. Acta*, 47:211-216.

Hidy, G.M., 1965. On the theory of coagulation of noninteracting particles in Brownian motion, *J. Coll. Sci.*, 20:123-144.

Holliday, L.M. and Liss, P.S., 1976. The behaviour of dissolved Fe, Mn and Zn in the Beaulieu estuary S. England, *Estuar. Coast. Mar. Sci.*, 4:349-353.

Hunter, K.A., 1983. On the estuarine mixing of dissolved substances in relation to colloid stability and surface properties, *Geochim. Cosmochim. Acta*, 47:467-473.

Hunter, K.A. and Leonard, M.L., 1988. The surface charge of suspended colloid stability and aggregation in estuaries: 1. Aggregation kinetics of riverine dissolved iron after mixing with seawater, *Geochim. Cosmochim. Acta*, 52:1123-1130.

Hunter, K.A. and Liss, P.S., 1979. The surface charge of suspended particles in estuarine and coastal waters, *Nature*, 282:823-825.

Hunter, K.A. and Liss, P.S. 1982. Organic matter and the surface charge of suspended particles in estuarine waters, *Limnol. Oceanogr.*, 27:322-335.

Hunter, K.A., Leonard, M.L., Carpenter, P.D. and Smith, J.D., 1990. The surface charge of suspended colloid stability and aggregation in estuaries: 2. The removal of dissolved iron during continuous mixing, in preparation.

Ives, K.J., 1978. Rate theories *in:* "The Scientific Basis of Flocculation", K.J. Ives, ed., Sijthoff and Njordhoff, The Netherlands, pp 3-38.

Matsunaga, K., Igarashi, K., Fukase, S. and Tsubota, H., 1984. Behaviour of organically bound iron in seawater of estuaries, *Estuar. Coast. Shelf Sci.* 18:615-622.

Mayer, L.M., 1982. Aggregation of colloidal iron during estuarine mixing: kinetics, mechanism and seasonality, *Geochim. Cosmochim. Acta*, 46:2527-2535.

Moore, R.M., Burton, J.D., Williams, P.J. LeB. and Young M.L., 1979. The behaviour of dissolved iron and manganese in estuarine mixing, *Geochim. Cosmochim. Acta*, 43:919-926.

Napper, D.H., 1977. Steric stabilisation, *Geochim. Cosmochim. Acta*, 58:390-407.

Ottewill, R.H. and Shaw, J.N., 1965. Stability of monodisperse polystyrene latex dispersions of various sizes, *Disc. Farad. Soc.*, 42:154-163.

Overbeek, J.Th.G., 1977. Recent developments in the understanding of colloid stability, *J. Coll. Interface Sci.*, 58:408-422.

Prieve, D.C. and Lin, M.M.J., 1982. The effect of a distribution in surface properties on colloid stability, *J. Coll. Interface Sci.*, 86:17-25.

Shapiro, J., 1964. Effect of yellow organic matter on iron and other metals in water, *J Amer. Waterworks Assoc.*, 56:1062-1082.

Sholkovitz, E.R., 1976. Flocculation of dissolved organic and inorganic matter during the mixing of river water and sea water, *Geochim. Cosmochim. Acta*, 40:831-845.

Sholkovitz, E.R., Boyle, E.A. and Price, N.B., 1978. The removal of dissolved humic acids and iron during estuarine mixing, *Earth Planet. Sci. Lett.*, 40:130-136.

Smith, J.D. and Longmore, A.R., 1980. Behaviour of phosphate in estuarine water, *Nature*, 287:532-534.

Windom, H.L., Beck, K.C. and Smith, R., 1971. Transportation of trace metals to the Atlantic Ocean by three southeastern rivers, *Southeast Geol.*, 1109-1181.

Smith, J.D. and Milne, P.J., 1981. The behaviour of iron in estuaries and its interaction with other components of the water, *in:* "River Inputs to Ocean Systems", United Nations, New York, pp 223-230.

# THE GENERATION OF SUSPENDED SEDIMENT

# IN RIVERS AND STREAMS

Brian L. Finlayson

Centre for Environmental Applied Hydrology
Department of Geography
University of Melbourne
Parkville, Victoria

## INTRODUCTION

The collection of adequate data on suspended sediment transport in rivers is, for the most part, confined to experimental and research activities, usually in relatively small catchments. By their very nature these are usually only for short time periods. In Australia there are no good records of suspended sediment transport available for any major river. There are a number of reasons for this state of affairs. Most suspended sediment transport occurs in association with the flood hydrograph and these are the most difficult events to monitor. No simple relationship exists between suspended sediment concentration and other measurable parameters such as discharge so that estimation of the transport rate for periods when no samples were taken is unreliable. Since the concentration of suspended sediment varies through the cross section of the flow, a single sample collected at one point does not necessarily yield a representative concentration. Measured concentration will also vary depending on the way in which a sample is collected so that specialised samplers are required. All these factors combine to make the collection of data which give a reliable measure of suspended sediment transport in the long term logistically difficult, time consuming and therefore expensive.

A knowledge and understanding of suspended sediment generation and transport is of interest in a wide range of disciplines; agriculture, zoology, reservoir management, geomorphology and aquatic chemistry to name but a few, and the problems posed by the highly discontinuous nature of suspended sediment transport in time and space are common to all of them. In this paper three aspects will be considered; the generation of suspended sediment in relation to the flood hydrograph, the sampling problem and the composition of the suspended load.

## SUSPENDED SEDIMENT AND THE FLOOD HYDROGRAPH

The velocities required to entrain suspended sediment are only reached by turbulent flow in open channels so that within a catchment the area contributing suspended sediment is confined to those parts of the catchment where such flows occur. The entrainment of sediment into the flow is assisted by raindrop impact which detaches smaller particles from otherwise stable aggregates thus facilitating erosion (Moss et al., 1979).

Flow of the required velocity occurs in stream channels at all stages of discharge though sediment transport is limited by the availability of suitable material and by bed armouring. At low discharge levels, during periods of baseflow, most streams transport

*Surface and Colloid Chemistry in Natural Waters and Water Treatment*
Edited by R. Beckett, Plenum Press, New York, 1990

sediment at rates well below their capacity. Increases in discharge associated with flood hydrographs are invariably accompanied by sharp increases in sediment concentration (Figure 1), usually from very low and stable preexisting levels. The increased depth of flow in the channel causes the inundation of those parts of the channel bed and banks where sediment is available for transport; during floods there is an extension of the actively flowing channel network (Blyth and Rodda, 1973) and surface runoff is also generated on the catchment slopes.

The generation of surface runoff outside the channel is a complex process. Two basic models are used to define runoff contributing areas. The first derives from the work of Horton (1933) where runoff occurs when rainfall intensity exceeds the infiltration capacity of the soil. This is known as infiltration excess overland flow and can, in theory at least, occur anywhere within the catchment. Conceptual models based on Horton's work have generally ignored the spatial variability he ascribed to this process and have assumed that runoff will be generated over the whole catchment at a rate determined by the excess of rainfall over catchment loss. While this approach is adequate for generalised runoff prediction models, it does not adequately reflect what really happens in most catchments and therefore has little value in defining source areas of sediment except perhaps in some areas of bare cultivated soils.

Experience over the past 20 years or so has shown that the partial contributing area model (Betson, 1964) provides a more realistic model of runoff generation, and also a tool for analysing source areas of suspended sediment (e.g. Finlayson, 1977). In this model runoff generation is governed by available storage capacity in the soil. When the soil moisture store is full, no further water can enter and additional rainfall becomes overland flow, irrespective of intensity. Partial contributing areas in a catchment consist of stream channels and areas of concave contour curvature where flow lines converge.

Recent work by O'Loughlin (1986) and others has shown that it is possible to predict partial contributing areas using a topographic model. The influence of many of the variables involved in runoff and sediment production, such as soil erodibility, rainfall

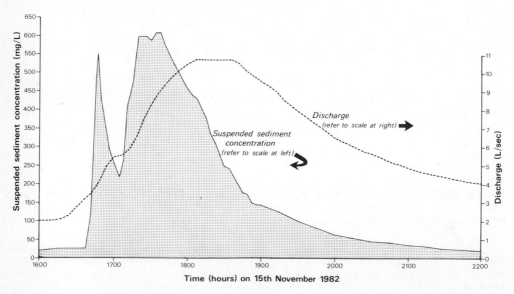

Figure 1.    Variations in suspended sediment concentration during the passage of a flood peak at Myrtle Creek No 1, Victoria. Sediment concentration derived from a continuous recording turbidity meter. Discharge data supplied by the Melbourne and Metropolitan Board of Works.

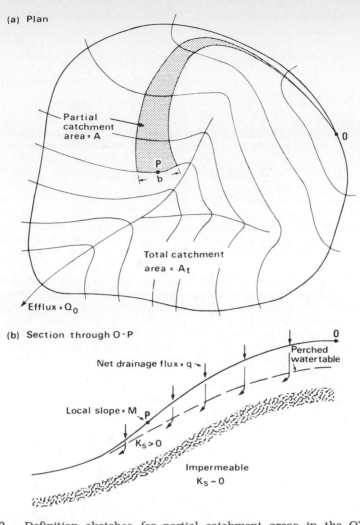

(a) Plan

Partial
catchment
area = A

0

P

b

Total catchment
area = $A_t$

Efflux = $Q_0$

(b) Section through O - P

0

Perched
water table

Net drainage flux = q

Local slope = M

P

$K_s > 0$

Impermeable
$K_s \sim 0$

Figure 2.    Definition sketches for partial catchment areas in the O'Loughlin
topographic model (Source: O'Loughlin, 1986).

erosivity, cover and slope, is known from plot studies.  Problems arise in extrapolating plot studies to whole catchments where topography becomes an important control. O'Loughlin's topographic model (Figure 2) enables real catchment topography to be incorporated into models of runoff generation and sediment production using either the partial area model or the infiltration excess model.  In the case of the partial area model, saturated areas are predicted for any given state of catchment wetness (Figure 3).  As can be seen from Figure 4, the size of the saturated area predicted by the model agrees well with actual runoff percentages from observed storms.

It is clear that where the partial area model applies (and this is probably the majority of cases) surface runoff and therefore suspended sediment will be generated from only a relatively small proportion of the catchment.  Using the same topographic model combined with the infiltration excess model of runoff generation Burch et al. (1986) have shown that local variations in topography lead to uneven distribution of sediment production (Figure 5) since both erosion and deposition occur locally.  Topographic analysis demonstrates the irregular distribution of sources of sediment production within the catchment and offers a tool for better catchment management.

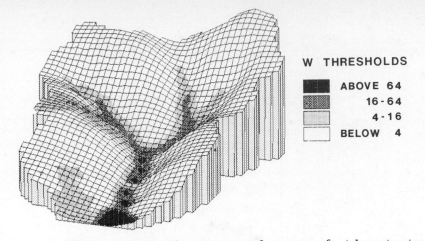

Figure 3.    Predicted zones of surface saturation for a range of catchment wetness values (W) with full forest cover and uniform drainage flux (Source: O'Loughlin, 1986).

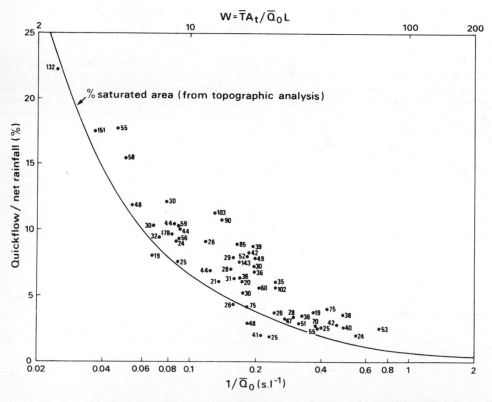

Figure 4.    Comparison of saturated areas predicted by the topographic model (W, top scale) and calculated from catchment runoff ($1/\bar{Q}_0$, bottom scale). Net rainfall in millimetres is shown for each storm (Source: O'Loughlin, 1986).

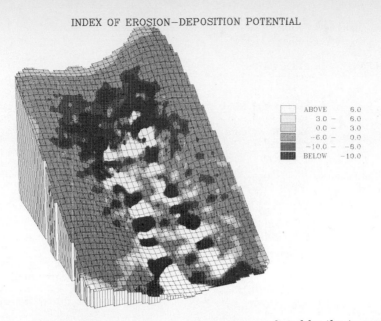

| | ABOVE | 6.0 |
| --- | --- | --- |
| | 3.0 – | 6.0 |
| | 0.0 – | 3.0 |
| | −6.0 – | 0.0 |
| | −10.0 – | −6.0 |
| | BELOW | −10.0 |

Figure 5.    Relative rates of erosion and deposition predicted by the topographic model.  Erosion is negative and deposition is positive (Source: Burch et al., 1986).

These two runoff generation models in fact represent end members of a continuum and in many catchments some runoff will be generated by both processes.  Ragan (1968) described a case where runoff was generated from partial contributing areas by low intensity rainfall and by infiltration excess during high intensity rainfall.  Clearly the predictability of both runoff generation and sediment production depends heavily on site-specific properties of individual catchments.

The work of Burch et al. (1986) demonstrates a phenomenon well known to fluvial geomorphologists, that of poor slope-channel coupling in sediment transport.  The occurrence of erosion on catchment slopes does not necessarily imply that the sediment eroded will be delivered to the main stream channel.  Thus the relationships between catchment treatments or landuse and sediment transport in trunk streams is often obscured by local slope storage effects as illustrated by the work of Ciesolka (1986) and Van Hooff and Jungerius (1984).

The storm hydrograph often includes a component of rapid throughflow in the catchment soils.  This occurs as turbulent flow through macropores and pipes in the soil and is capable of transporting suspended solids.  A review of pipeflow processes is provided by Jones (1971) while Pilgrim and Huff (1983) and Imeson et al. (1984) report examples of sediment transport through soil macropores.

In large streams much of the sediment being transported originates from the channel bed and banks.  This is probably also the case for small streams as shown by Duijsings (1984) who documented the various source areas in a small forested catchment (Table 1).  Loughran et al. (1986), working in a partially cultivated catchment, calculated that 93% of the sediment was derived from vineyards occupying only 60% of the catchment area.  The remaining 7% came from forest and grazing land.  Within forested areas, Anderson (1974) has shown that forest roads near streams were the single greatest contributors of sediment.

A number of methods have been proposed for identifying source areas of sediment using characteristics of the sediment itself.  These include sediment mineralogy (Wall and

Table 1. Sediment Budget for the Schrondweilerbach Catchment in Luxembourg (Source: Duijsings, 1984)

| INPUT/OUTPUT SOURCE | PROCESS | LOAD (kg/2 yrs) | % |
|---|---|---|---|
| stream | lateral corrasion | 20650 | 22.9 |
| banks | subsoil fall | 18860 | 21.0 |
| | mass movements | 6850 | 7.6 |
| | rainsplash | 1240 | 1.4 |
| | soil creep | 400 | 0.4 |
| | BANK TOTAL | 48000 | 53.3 |
| valley | rainsplash & | | |
| slopes | overland flow | 38000 | 42.2 |
| | throughflow | 4000 | 4.5 |
| | SLOPES TOTAL | 42000 | 46.7 |
| | TOTAL INPUT | 90000 | 100.0 |
| instream | suspended load | 86300 | 92.8 |
| transport | bed load | 6700 | 7.2 |
| | TOTAL OUTPUT | 93000 | 100.0 |

Wilding, 1976), radioactivity (Ritchie and McHenry, 1975), magnetic properties (Oldfield et al., 1979), pollen (Brown, 1985) and Caesium 137 (Loughran et al., 1982). Many of these techniques are experimental and all depend to some extent on there being significant variation in the property of interest around the catchment.

**THE SAMPLING PROBLEM**

The sampling problem in suspended sediment studies can be subdivided into two components. The first concerns the distribution of sample collection through time so as to adequately characterize the rate of sediment transport over some period of interest. The second concerns the manner of collection of a single sample so as to produce a value for mean concentration through the cross section of the flow. There is a third problem, a subset of the second, which concerns the collection of sufficient sediment for reliable analyses of composition, particularly of trace metals and nutrients.

The processes of runoff generation described above lead to highly variable rates of sediment production between catchments and for any individual catchment from time to time. This situation is further complicated by sediment exhaustion effects as illustrated in Figure 1. Sediment available to be transported in the runoff generating area is removed during the initial rise of the hydrograph and the peak of sediment concentration is reached before the discharge peak. On the recession limb of the hydrograph the concentration for any given discharge is much lower than on the rising limb. This hysteresis is evident when concentration is plotted against discharge and the individual points joined in time sequence. When hysteresis loops are plotted for a series of storms from the same catchment (Figure 6) the reasons for the poor statistical relationship between concentration and discharge become obvious. Clockwise hysteresis is the most common type but anticlockwise hysteresis has also been observed. Klein (1984) suggests that clockwise hysteresis occurs when sediment is derived from the bed and banks and areas adjacent to the channel while anticlockwise hysteresis occurs when the sediment

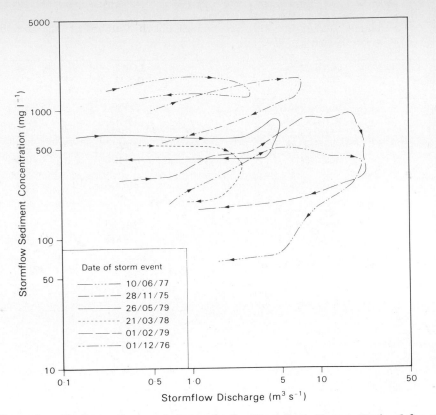

Figure 6.    Clockwise hysteresis loops for the River Dart, Devon, England for six
             storm peaks over the period 1975-79 (Source:  Walling and Webb,
             1982).

(and the runoff) source area is on the upper part of the catchment slopes. This behaviour
can therefore be related to the mechanism of runoff generation: clockwise hysteresis for
runoff from partial source areas and anticlockwise hysteresis for infiltration excess
runoff.

The poor statistical relationship between concentration and discharge means that
extrapolation techniques such as the rating curve method are inappropriate for
calculating total load and the limitations of these methods have been well documented
(see, for example, Walling, 1977).  The most reliable way to measure total sediment load
over any given period is to sample all the flood peaks.  This can be done manually or with
automatic samplers triggered by rise in stage.  In either case the sampling and sample
analysis is time consuming and expensive.

An alternative approach, which somewhat surprisingly has been little used, is to
record concentration continuously using either turbidimetry or nephelometry.  Walling
(1977) and Walling and Webb (1982) have used continuous turbidity records, calibrated
with gravimetric analysis of samples, as a standard against which to compare
interpolation and extrapolation methods for determining total load.  Finlayson (1985) has
described a method for the field calibration of turbidity meters and his calibration curve
(Figure 7), based on 132 samples collected from storm runoff over a twelve month period,
explains 97% of the variance in suspended solids concentration.  Gilvear and Petts (1985)
found the relationship between turbidity and suspended solids in the Afon Tryweryn
(Wales) to be quite variable due to changes during the passage of the flood peak in size,
configuration and composition of the material being transported. They suggest that
separate ratings may be needed for the rising and falling limbs of the hydrograph.

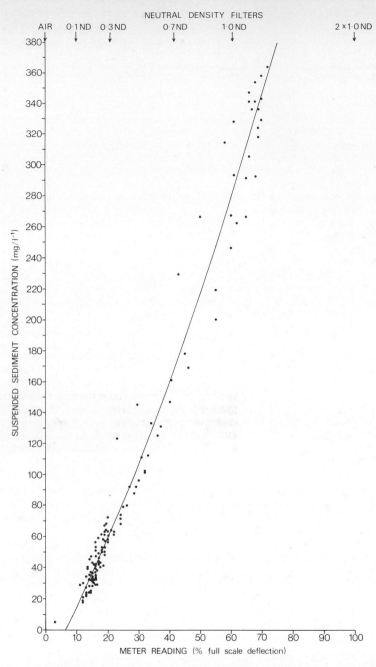

Figure 7. Calibration curve for suspended sediment concentration and turbidity for Myrtle Creek No 1, Victoria (Source: Finlayson, 1985).

The applicability of turbidity as a means of measuring suspended sediment concentration has been questioned by Oades (1982). He points out that theory predicts a poor correlation and quotes results obtained by Croome (1980) for the River Murray at Albury-Wodonga in support. In that case turbidity explained only 44% of the variance in suspended solids concentration. In rivers like the Murray with abundant fine clay in the suspended load the use of turbidity may not be appropriate but there is strong evidence to suggest that in many rivers it may work sufficiently well to make it the most cost-effective means of collecting data for total load, particularly if this information is required over a long period. Continuous recording of turbidity is incorporated into the national water quality monitoring programmes in some countries such as Finland (Kohoven, 1984) and the United States (Herricks, 1984). At present its use in Australia is confined to a few experimental sites.

While frequent sampling of flood peaks or the use of recording turbidity meters can help overcome the problem of temporal variability in suspended sediment concentration, these methods do not address the second sampling problem of how to collect a representative sample (or samples) from the cross section of the flow. For small streams of the type frequently used as experimental catchments, the cross sectional area of the flow is so small that it is adequately represented by a single point sample. In large channels this is not the case. Particles in suspension have a tendency to settle to the bed under the influence of gravity at a rate controlled by particle size, shape and density and the viscosity of the water. In turbulent flow this tendency to settle out is balanced by upwards movements in the flow and an equilibrium concentration profile is established. Both concentration and mean grain size increase towards the bed as shown in Figure 8. Similarly there is lateral variation in concentration and grain size distribution (Figure 8).

The lateral and vertical distribution of velocity is also shown in Figure 8. This velocity pattern explains some of the concentration variation and is a major factor in determining the distribution of suspended solids discharge through the cross section. This distribution of particles in suspension and velocity of flow must be taken into account during sample collection.

Samplers should be designed to fulfil three requirements: flow velocity at the sampler intake should be the same as the flow velocity in the stream, the presence of the sampler should cause a minimum of disturbance to the flow, and the sampler intake should be oriented directly into the flow in both the vertical and horizontal planes (Gregory and Walling, 1973). Commercial point and depth integrating samplers meet these requirements and are used in the field to collect discharge-weighted samples. The ideal sampling strategy would be to develop a relationship between discharge weighted concentration for the cross section and concentration at a single point, though to date no such results have been reported in the literature.

Conventional samplers collect relatively small volumes of sample (ca. 500 mL) and in most cases the nonfilterable residue is in milligram quantities. Where the purpose of the exercise is to determine total concentration, such samples are entirely adequate. Where detailed analysis of the composition of the sediment is intended, much larger samples are needed. The most efficient solution to this problem is to separate sediment from water on-site using a continuous flow centrifuge as reported by Hart (1987) in a study of trace metal and radionuclide transport in Magela Creek, NT.

Sampling for suspended sediment is a complex operation and, not surprisingly, much of the work reported in the literature is based on an inadequate sampling design. Frequently the methods of sample collection are not specified despite their importance in determining the reliability of the results.

## THE COMPOSITION OF SUSPENDED SEDIMENT

The suspended particulate load of streams consists of a largely "natural" component of inorganic mineral matter either as discrete grains or aggregates and living and dead organic matter to which may be added a wide range of artificially derived material depending on the source area of the flow. The upper size limit of the suspended fraction is determined by velocity and discharge and will therefore vary from time to time. The lower size limit is more difficult to define since there is no clear size distinction between a fine colloid, a macromolecule and true solution (Oades, 1982). Separation of "suspended" from "dissolved" is usually carried out using 0.45 μm filters for purely practical reasons since filters of this pore size represent an acceptable compromise between time taken for samples to pass through the filters and the recovery of the bulk of the solid particles. For many applications of suspended sediment analysis precise definition of the lower size limit is immaterial since the weight of the sample is dominated by the larger grain sizes. However, where the sorption capacity is being investigated, the finer fraction is of overriding importance and separation at a lower size limit is justified.

Walling and Moorehead (1987) have recently reviewed the literature on the particle size characteristics of suspended sediment and point out that despite the importance of particle size in understanding environmental processes, our knowledge of spatial and

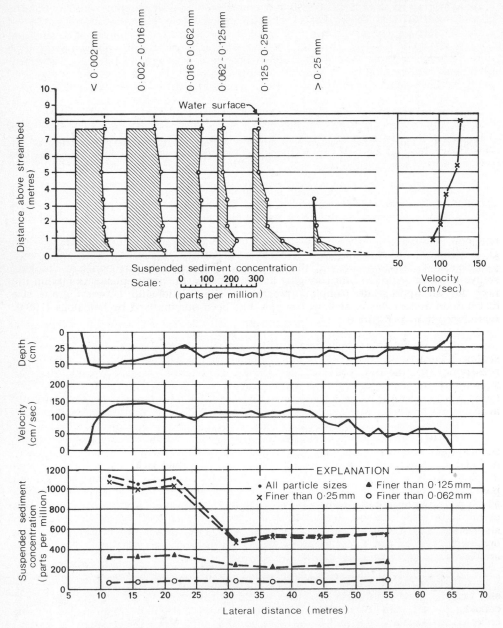

Figure 8. The distribution of concentration of various particle size ranges and velocity in a stream cross section (Source: Nordin and Richardson, 1971).

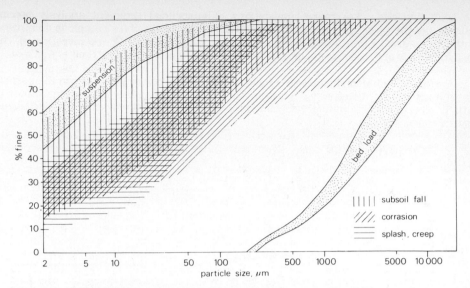

Figure 9.    Cumulative grainsize frequency zones for the different modes of sediment supply and transport in the Schrondweilerbach, Luxembourg (Source: Duijsings, 1984).

temporal variability is somewhat limited.  Their review points to the fact that there is great variability within and between rivers at the global scale and at the local scale.  Since particle size distribution may vary through the flood hydrograph in response to changing source areas of sediment and changing flow conditions, further emphasis is laid on the need for an appropriate sampling program.    The relationship between grain size distribution and modes of sediment supply has been summarised by Duijsings (1984), reproduced here as Figure 9.

It is difficult to apply conventional particle sizing techniques to suspended sediment because of the small sample volume usually available and the difficulty of preserving the natural state of aggregation throughout sampling and analysis. Microscopic, optical and electronic methods have been reviewed by Swift et al. (1972). Laser particle sizing apparatus (Butters and Wheatley, 1982) is the best method currently available but has been little used in this context as yet.

The need to distinguish between the mineral and organic fractions of suspended sediment has long been recognized by geomorphologists interested in determining rates of denudation.  Suspended organic matter is also important biologically as a source of food for fish, as a pollutant, and since it is partially composed of plant and animal populations.  Few generalisations can be made about the proportion of suspended matter which is organic in origin but there is no doubt that it represents a significant proportion of the total load in most cases.  In a detailed study of the organic component of suspended sediment in catchments in North Yorkshire, Arnett (1978) found that the organic fraction ranged from 40% during baseflow to 20% at high flows and these figures are consistent with others reported in the literature.

Few results have been reported of the distribution of organic matter across the size ranges.  Hart (1986) has used loss on ignition to measure the organic fraction of colloidal matter, fine particulate matter and coarse particulate matter in samples collected in Magela Creek and found it to be 83%, 60% and 21% respectively.  These percentages are much higher than in the catchment soils demonstrating both the selectivity of the erosion process and the importance of in-stream sources of organic matter (Gilvear and Petts, 1985).

The organic fraction of particulates in urban runoff also appears to be concentrated in the finer fraction. Ellis (1979) found that 15-70% of the organic component of suspended solids in runoff from north-west London were finer than 0.06 mm. The sampling problems discussed earlier also apply in urban catchments. Weeks (1982) reported an initial flush effect for particulates in urban runoff as well as clear distinctions in composition and total load between residential and industrial areas.

## CONCLUSIONS

Three main points arise from this brief overview of suspended sediment production:

1.  An understanding of the mechanisms of suspended sediment generation and source areas within catchments, be they urban or rural, must be based on an appropriate and realistic model of runoff generation. The variability of catchment behaviour in terms of runoff generation is such that although the basic processes are understood and can be modelled, the transportability of models is limited by the need to recalibrate on each catchment. This process can be facilitated by topographic analysis. Source areas of suspended sediment are highly localised within catchments and this fact should be understood and taken account of in the development of effective catchment management programmes. More research is needed on the integration of sediment production into runoff generation models.

2.  Sampling remains an unresolved problem in suspended sediment studies. At present, effective sampling strategies are confined largely to experimental and research situations. There is a clear need for more research into cost-effective sampling methods for operational use.

3.  Suspended sediment transport in Australian rivers is poorly known even at the simplest level of total loads. The important role suspended sediment plays in pollutant and nutrient transport indicates a need for greater effort in the analysis of its composition and its role in environmental processes.

## REFERENCES

Anderson, H.W., 1974. Sediment deposition in reservoirs associated with rural roads, forest fires, and catchment attributes, *IASH - AISH Publ.*, 113:87-95.

Arnett, R.R., 1978. Regional disparities in the denudation rate of organic sediments, *Zeit. fur Geomorph. Suppl.*, 29:169-179.

Betson, R.P., 1964. What is watershed runoff?, *J. Geophys. Res.*, 69:1541-1551.

Blyth, K. and Rodda, J.C., 1973. A stream length study, *Water Resources Res.*, 9(5):1454-1460.

Brown, A.G., 1985. The potential use of pollen in the identification of suspended sediment sources, *Earth Surf. Proc. and Landforms*, 10:27'32.

Burch, G.J., More, I.D., Barnes, C.J., Aveyard, J.M. and Barker, P.J., 1986. Modelling erosion hazards: a total catchment approach, *in:* "Hydrology and Water Resources Symposium", Brisbane, November 1986, The Institution of Engineers, Australia, National Conference Publication No 86/13, pp 345-349.

Butters, G., and Wheatley, A.L., 1982. Experience with the Malvern ST 1800 laser diffraction particle sizer, *in:* "Particle Size Analysis 1981", N.G. Stanley-Wood, and T. Allen, eds., Wiley-Heyden, London.

Ciesolka, C., 1986. "Land Degradation in the Nogoa Watershed, Central Queensland", Unpublished Research Report, Department of Primary Industries, Queensland.

Croome, R., 1980. "Turbidity, colour, nutrients and plankton of the River Murray", Unpublished Report to the River Murray Commission Water Quality Committee, Albury.

Duijsings, J.J.H.M., 1984. "Streambank Contribution to the Sediment Budget of a Forest Stream", Publication No 40, Laboratory for Physical Geography and Soil Science, University of Amsterdam, Amsterdam.

Ellis, J.B., 1979. The nature and sources of urban suspended sediments and their relation to water quality: a case study from north-west London, *in:* "Man's Impact on the Hydrological Cycle in the United Kingdom", G.E. Hollis, ed., Geo Books, Norwich.

Finlayson, B.L., 1977. "Runoff Contributing Areas and Erosion", Research Paper 18, School of Geography, Oxford University.

Finlayson, B.L., 1985. Field calibration of a recording turbidity meter, *Catena,* 12:141-147.

Gilvear, D.J. and Petts, G.E., 1985. Turbidity and suspended solids variations downstream of a regulating reservoir, *Earth Surf. Proc. and Landforms,* 10:363-373.

Gregory K.J. and Walling, D.E., 1973. "Drainage Basin Form and Process", Edward Arnold, London.

Hart, B.T., 1986. Transport and fate of trace metals and radionuclides, *in:* "Annual Research Report 1985/86", Water Studies Centre, Chisholm Institute of Technology, Melbourne.

Hart, B.T., 1987. Fate of trace metals and radionuclides in the Magela Creek system, northern Australia, *in:* "Paper presented to the 9th Analytical Chemistry Conference", Sydney, April, Royal Australian Chemistry Institute.

Herricks, E.E., 1984. Aspects of monitoring in river basin management, *Water. Sci. and Technol.,* 16:259-274.

Horton, R.E., 1933. The role of infiltration in the hydrological cycle, *Trans. Amer. Geophys. Union,* 14:446-460.

Imeson, A.C., Vis, M. and Duijsings, J.J.H.M., 1984. Surface and subsurface sources of suspended solids in forested drainage basins in the Keuper region of Luxembourg, *in:* "Catchment Experiments in Fluvial Geomorphology", T.P. Burt and D.E.Walling, eds., Geo Books, Norwich.

Jones, A., 1971. Soil piping and stream channel initiation, *Water Resources Res.,* 7:602-610.

Klein, M., 1984. Anticlockwise hysteresis in suspended sediment concentration during individual storms: Holbeck Catchment, Yorkshire, England, *Catena,* 11:251-257.

Kohoven, T., 1984. Automatic monitoring of river water quality, *Water. Sci. and Technol.,* 16:289-294.

Loughran, R.J., Campbell, B.L. and Elliott, G.L., 1982. The identification and quantification of sediment sources using Cs137, *IAHS - AISH Publ.,* 137:361-369.

Loughran, R.J., Campbell, B.L. and Elliott, G.L., 1986. Sediment dynamics in a partially cultivated catchment in New South Wales, Australia, *J. Hydrol.,* 83:285-297.

Moss, A.J., Walker, P.H. and Hutka, J., 1979. Raindrop stimulated transportation in shallow water flows: an experimental study, *Sed. Geol.,* 22:165-184.

Nordin, C.F. and Richardson, E.V., 1971. Instrumentation and measuring techniques, *in:* "River Mechanics 1", H.W.Shen, ed., H.W.Shen, Fort Collins, Colorado, USA.

Oades, J.M., 1982. Colour and turbidity in water, *in:* "Prediction in Water Quality", E.M. O'Loughlin and P. Cullen, eds., Australian Academy of Science, Canberra.

Oldfield, F., Rummert, T.A., Thompson, R. and Walling, D.E., 1979. Identification of suspended sediment sources by means of magnetic measurements: some preliminary results, *Water Resources Res.,* 15:211-218.

O'Loughlin, E.M., 1986, Prediction of surface saturation zones in natural catchments by topographic analysis, *Water Resources Res.,* 22(5):794-804.

Pilgrim, D.H. and Huff, D.D., 1983. Suspended sediment in rapid subsurface stormflow on a large field plot, *Earth Surf. Proc. and Landforms,* 8:451-463.

Swift, D.J.P., Schubel, J.R. and Sheldon, R.W., 1972. Size analysis of fine grained suspended sediments: a review, *J. Sed. Pet.,* 42(1):122-134.

van Hooff, P.P.M. and Jungerius, P.D., 1984. Sediment source and storage in small watersheds on the Keuper Marls in Luxembourg, *Catena,* 11:133-144.

Wall, G.J. and Wilding, L.P., 1976. Mineralogy and related parameters of fluvial suspended sediments in northwestern Ohio, *J. Environ. Qual.,* 5:168-173.

Walling, D.E., 1977. Limitations of the rating curve technique for estimating suspended sediment loads, with particular reference to British rivers, *IAHS - AISH Publ.,* 122:34-48.

Walling, D.E. and Moorehead, P.W., 1989. The particle size characteristics of fluvial suspended sediment: an overview, *Hydrobiologia*, 176/177:125-149.

Walling, D.E. and Webb, B.W., 1982. Sediment availability and the prediction of storm-period sediment yields, *IAHS - AISH Publ.*, 137:327-337.

Weeks, C.R., 1982, Pollution in Urban Runoff, *in:* "Water Quality Management-Monitoring Programs and Diffuse Runoff", B.T. Hart, ed., Chisholm Institute of Technology, Melbourne.

# APPLICATION OF THE URANIUM DECAY SERIES

## TO A STUDY OF GROUND WATER COLLOIDS

Richard T. Lowson and Stephen A. Short

Australian Nuclear Science and Technology Organisation
Lucas Heights Research Laboratories
Lucas Heights, New South Wales

## INTRODUCTION

In recent years, the study of surface and groundwater colloids has gained considerable momentum for a number of reasons. These include the failure of computer models to account correctly for the observed mass transport, the concern of environmental regulating organisations that not all sources of these colloids had been accounted for, and the development of new techniques for the collection and study of very dilute colloid systems.

A major problem is that colloids are usually present in very dilute concentrations, making detection and validation difficult. Concentration by factors of up to 100 may still leave the colloid at below the level of element detection by atomic absorption or inductively coupled plasma spectroscopy. It may be possible to carry out neutron activation analysis on the sample but the colloids cannot be examined by point of zero charge, isoelectric point, scanning electron microscopy analysis, etc.. Further concentration by factors of up to 10 000 may alter the structure of the colloid to the point of irrelevance.

Concern has been expressed that colloids may be the transport medium of insoluble radionuclides such as natural thorium isotopes and man-made plutonium isotopes. Although these radionuclides would be present at very low mass concentrations, the specific activity of the radionuclide may give significance to the activity concentration. Thus computer models based solely on solution/precipitation chemistry and designed to predict mass transport of these materials from buried waste may be flawed, and there is a need to develop methods to quantify the colloid contribution.

In the high concentration regime, colloids may be separated according to size, specific gravity or charge. However, the very low concentration in ground waters restricts the techniques applicable to those based on sized so evidence for the existence of a colloid is by inference rather than by direct observation. Possible techniques are gel filtration, centrifugation, dialysis and ultrafiltration, and have been reviewed by Salbu (1984) and by Benes and Majer (1980). Work at Lucas Heights has concentrated on the use of the hollow fibre ultrafiltration membrane, a technique which allows sampling of large amounts of water (>1000 L) with sampling time being the only limitation.

After a single pass through an ultrafiltration membrane, and with a concentrating factor of around 50, the elemental concentration may still be beyond the limits of atomic absorption. However, radioisotopes of the natural uranium and thorium decay chains.

*Surface and Colloid Chemistry in Natural Waters and Water Treatment*
Edited by R. Beckett, Plenum Press, New York, 1990

Table 1. Classification of Suspended Matter

| Name | Phase | Diameter (nm) | MW (Dalton) | Sedimentation Rate | Optical Properties |
|------|-------|---------------|-------------|--------------------|--------------------|
| pebble | suspension | $(4\text{-}54)\times10^6$ | $>10^8$ | 35-120 cm/s | visible |
| gravel | suspension | $(2\text{-}4)\times10^6$ | $>10^8$ | 20-35 cm/s | visible |
| sand | suspension | $(62\text{-}2000)\times10^3$ | $>10^8$ | 0.4-20 cm/s | visible |
| silt | colloid | $(0.4\text{-}62)\times10^3$ | $>10^8$ | 0.5-124 km/y | optical microscope |
| clay | colloid | $(2\text{-}4)\times10^3$ | $>10^8$ | 0.1-0.5 km/y | optical microscope |
| clay | colloid | 1000 | $10^8$ | $<0.1$ km/y | optical microscope |
| clay | colloid | 450 | $10^7$ | $<0.1$ km/y | ultramicroscope |
| clay | colloid | 100 | $10^6$ | $<0.1$ km/y | ultramicroscope |
| clay | colloid | 10 | $10^4$ | $<0.1$ km/y | not visible |
| clay | true solution | 1 | $10^2$ | $<0.1$ km/y | not visible |

Compiled from Yaris and Cross (1979) and Salbu et al. (1984).

can be detected at ultra low mass concentrations by determining the alpha activity. This allows information about a colloid to be obtained by observing the entrained radioactivity in the ultrafiltration and colloid concentrate fractions sampled down-gradient of uranium ore bodies.

## The Colloid Phase

The boundaries between true solution, colloid and suspension are somewhat arbitrary. Classification may be based on molecular weight, particle size, settling rate, limits on separation technique and methods of visual observation. The classifications are not directly interchangeable because density and shape vary from material to material. Table 1 gives a guide to the classification of suspended matter but it should not be considered exhaustive and the nomenclature used by different authors has varied considerably.

A further problem was the introduction of the unnecessary terms radiocolloid, true or real colloid, pseudocolloid and types I and II colloids. Liesner et al. (1986) attributed the term radiocolloid to Paneth (1913a,b), who used it to describe colloids containing a radionuclide. Kepak (1971) reviewed the colloidal properties of radioactive elements and retained the term. The term was derived from the method of detection (alpha, beta or gamma counting) and the ability to detect a radionuclide at mass concentrations well below chemical methods of detection. However, radioactivity is a property of the nucleus not of the atom; colloid formation is a property of the atom and not of the nucleus. Indeed a phase and its definition should in fact be independent of the method of detection.

The terms real or true colloid, pseudocolloid, and types I and II colloids arose from a need to distinguish between colloids of pure compounds such as colloidal hydroxides and colloids with a sorbed component, where the sorbed component was used to detect the colloid. The German literature refers to these materials as Eigenkolloide (self colloid) and Fremdkolloide (foreign colloid). This type of classification is unnecessary and should be abandoned unless, perhaps, it is shown that the process of sorption radically changes the properties of the colloid.  If such is the case, the colloid should be redefined in of its elemental composition rather than given an inappropiate generic classification.

## Ultrafiltration

Ultrafiltration, is a pressure driven technique, based on the forced flow of solvent through a porous membrane while retaining the concentrated solute/colloid.  Separation depends on the relationship between the size of the particle and the size of the pores.  The first ultrafiltration membranes were made from cellulose acetate, a material which is still extensively used.  There is now a trend to use more inert polymers such as polysuphone. The membrane is laid down on a support, which may be either a flat sheet, a spirally wound sheet or a tube.  The tubular support has the advantage that fouling of the inlet side is minimised by the tangential flushing action of the solution passing down the tube. The membrane should be asymmetric; that is to say the pore diameter should expand from the inlet side to the outlet side to eliminate internal clogging of the pores.

The system used in these laboratories is an Amicon type H10P100-20 asymetric hollow fibre ultrafiltration unit with a nominal molecular weight (MW) cut-off point at 100,000 Dalton and an effective surface area of 0.834 m$^2$.  The unit consists of a bundle of 1000 hollow fibre tubes (o.d. 1 mm, i.d. 0.5 mm, length 63.8 cm), sealed into an assembly support so that the supply water is directed to the inside of the tubes.  The membrane material is a dense skin of polysulphone, 0.1 to 1.5 μm thick, mounted on a much thicker (50 to 250 μm) spongy open cell layer of the same polymer.  The solvent, in this case water, passes through the membrane to produce an ultrafiltrate product and the concentrated solute (colloid concentrate) is collected at the exit manifold.

Although ultrafiltration membranes are characterized by a nominal molecular weight cut-off, the implication that the cutoff is sharp is incorrect.  Studies by Fane et al. (1981) have shown that the pore size has a Gaussian distribution across a range; they concluded that 50 per cent of the solvent flow is through 20 to 25 per cent of the pores. The quoted cutoff is determined using polymers of known molecular weight.  Different manufacturers use different polymers such as yellow dextran, cytochrome C, proteins, and polyethylene glycol, for characterization of the membrane.  Retention is not a simple function of molecular weight.  Spherical proteins may be retained more than linear molecules of the same molecular weight, so the method of measurement should be always stated (Cooper, 1984).

If an ultrafiltration membrane is considered as a collection of capillary pores through which Poiseuille flow occurs, then the solvent flux, $J_u$ of the ultrafiltrate across the membrane is given by

$$J_u = \frac{K' \Delta P}{nh} \qquad (1)$$

where K' is the hydraulic permeability, n is the solvent viscosity, $\Delta P$ is the hydrostatic pressure difference across the membrane, and h is the membrane thickness.

If C is the colloid concentratation in the feed stream, $C_u$ is the colloid concentration in the ultrafiltrate and $C_{cc}$ is the colloid concentration in the colloid concentrate, the colloid flux, $\tau_u$, across the membrane is given by

$$\tau_u = J_u C_u \qquad (2)$$

Defining a rejection coefficient $\sigma$ as

$$\sigma = 1 - C_u / C_{cc} \qquad (3)$$

equation (3) will arrange to

$$\tau_u = C_{cc}(1-\sigma)J_u \qquad (4)$$

The term $(1-\sigma)$ refers to the fraction of solvent flux passing through pores large enough to allow the transport of colloid and $\sigma$ is the fraction of solvent flux passing through pores too small to permit the transport of colloid.

In the absence of build-up of material on the membrane surface, the flux is a linear function of the hydraulic pressure and the rejection coefficient is a pressure independent constant. In practice, however, material becomes trapped on the surface. This material, referred to as the gel layer, will alter the flow and separation characteristics of the membrane. Normally it is in a state of dynamic equilibrium between formation and breakdown. The rate of formation of the gel layer is given as the product of the solvent flux of the ultrafiltrate and the bulk concentration, $J_uC$. Film breakdown due to back-diffusion into the bulk liquid is given by the product $D(dC/dx)$ so that at steady state conditions

$$J_uC = D(dC/dx) \qquad (5)$$

where $dC/dx$ is the concentration gradient across the gel layer and $D$ is the diffusion coefficient. For a fixed concentration $C_o$ at the membrane surface, equation (5) integrates to

$$J_u = K \ln(C_o/C) \qquad (6)$$

where $K$ is the mass transfer coefficient.

This general relationship is frequently observed. It indicates that the flux of ultrafiltered water is controlled by the transfer of species to and from the gel layer. The gel layer will break down under increased velocities past the membrane surface or increased temperatures, thus increasing the flux of ultrafiltered water. Increased pressure across the membrane will promote formation of the gel layer and reduce the flux of ultrafiltered water. In the extreme case, the solvent flux is independent of pressure and the membrane is said to be gel polarised.

The gel layer not only decreases the flux of ultrafiltrate, but it also increases the flux of colloid across the membrane owing to the increased concentration of colloid at the membrane surface, leading to a lower rejection coefficient.

Optimal operation requires a high tangential surface velocity and low cross membrane pressures. Unfortunately the concentration of colloids in ground water in our experiments was so low that the system had to operated with a maximum water flux through the membrane and a minimum solution flux past it in order to concentrate the colloid up to observable levels. Consequently the unit was operating under the worst possible condition for gel layer formation. However, it was considered that even under these conditions the colloid concentration was too low for the gel layer to have a significant effect on the rejection coefficient.

The general formula for the concentration factor is given by

$$C/C^o = (V_{cc}/V^o{}_{cc})^{-\tau} \qquad (7)$$

where $V_{cc}$ is the volume of the colloid concentrate and the superscript $o$ signifies the initial conditions. The concentration factor, $C/C^o$, is controlled solely by the operating parameters.

Figure 1 is a diagram of the unit used in the field. A stainless steel submersible Grundfos electric pump was lowered into the bore and set above the point of recharge. The water was pumped at a rate slow enough to avoid excessive draw-down and contamination by atmospheric oxygen. At least three bore volumes were pumped out before sampling commenced. The water was passed successively through 5 μm and 1 μm cartridge filters before passing to the ultrafiltration unit. The unit was operated in single pass mode with the feed stream entering the unit at around 170 kPa (25 psi), at a flow rate of between 6 to 10 L/min. The colloid product stream was throttled back to 0.1 L/min to give a concentration factor of between 50 to 100, depending on the feed rate.

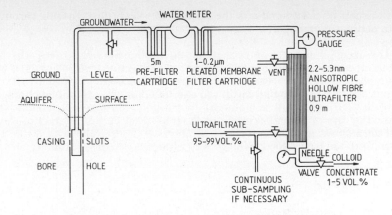

Figure 1.    Ground water colloid sampling system.

The ultrafiltrate, which was up to 99 per cent of the feed stream, was subsampled at the same rate as the colloid concentrate.  A 20 L sample of each fraction was collected.  The fractions were acidified to about pH 1 with concentrated HCl and sent to Lucas Heights for analysis.

The alternative method of sampling is successive recirculation.    There is an inherent problem with recirculation in that the colloid may undergo significant modifications as the concentration is increased by factors in excess of one thousand.  The modified colloid would no longer be representative of the ground water colloid and may be a strong adsorber of solute molecules, leading to erroneous interpretation.

**The Uranium and Thorium Decay Series**

Figure 2 illustrates the upper members of the three natural uranium and thorium decay series.   In a closed system and for periods in excess of 1.2 million years, the uranium decay chain will be in a state of secular equilibrium.   That is to say, the decay rate of the daughter is equal to the ingrowth rate from the parent.   In contrast, in open

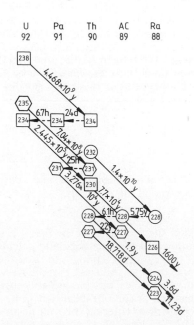

Figure 2.    The upper members of the uranium and thorium decay series.

systems fractionation occurs through a variety of mechanisms and the observed disequilibria provides a means of measuring the kinetics of a number of natural processes which have occurred in the last 1 million years.

The isotopic concentration is determined by adding a man-made isotope such as $^{236}U$ or $^{232}U$ which acts as an isotopic yield tracer. The element of interest is separated out by electodeposition as an infinitely thin source on a stainless steel planchette and its alpha spectrum determined with a high resolution alpha spectrometer. The background for alpha counting is very low, allowing a practical detection limit to be set at 1 count per minute, which depending on the efficiency of the counter, is equivalent to about 1 µg $^{238}U$. The method has been described in detail by Lowson and Short (1986).

## RESULTS AND DISCUSSION

### Sampling Sites

Samples were collected down gradient from the Nabarlek and Koongarra uranium ore bodies in the Alligator Rivers Uranium Province of the Northern Territory of Australia. The region is characterized by a summer monsoonal climate with well defined wet and dry seasons. The average rainfall is $1500 \pm 300$ mm and falls only in the wet season which is from December to March. In the dry season very little rain falls. The temperatures vary from around 36°C in October to about 30°C in March. The annual evaporation from surface waters is approximately 2200 mm.

The Nabarlek ore body, (Figure 3), is part of a series of Lower Proterozoic sediments laid down between 1800 to 1900 million years ago. The rocks comprise of schistose quartzites, quartz mica schists, muscovite schists and amphibole schists with common biotite, feldspar and garnet. The rocks are affected by later alteration.

The ore body was contained within massive fine grained chloritic rock. It had surface expression, struck north north west and dipped 30° to 40° to the east. It had a tabular shape, was 230 m long and had an average thickness of 10 m with wide fluctuations. The mineralisation extended to a known maximum depth of 84 m, with most of the ore being contained in the interval between the surface and 45 m vertical depth. The ore body was mined in its entirety in 1978. The Nabarlek open cut is now used to store the tailings. This would not have affected the sampling of waters down gradient of the ore body in the time between mining and sampling.

The depth of the standing water table in the bores varied from 1 to 18 m and reflected the variable permeabilities of the rock. Measurements indicated an easterly flow during the wet season with gradients of 1 in 16 to 1 in 40, controlled mainly by the

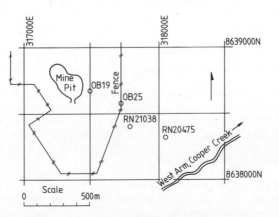

Figure 3.    The Nabarlek sampling site.

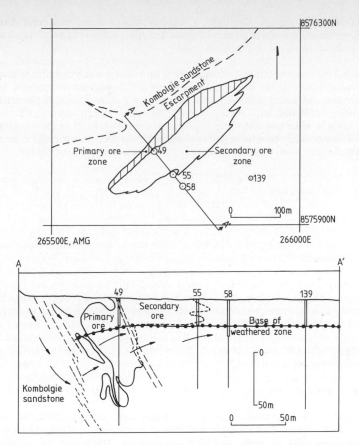

Figure 4.    The Koongarra sampling site (a) plan view (b) in section.

topography.    During the dry season, the flows changed to south westerly possibly controlled by bedrock fault structures. The measurements suggested that during the wet season, a perched water table occurs in the near-surface sands and ferricretes which overlay the weathered schist bedrock. Four cased and slotted bores were selected in an approximate straight line down the mean hydraulic gradient from the deposit.

The Koongarra ore body is geologically similar to that at Nabarlek, (see Figures 4a,b).    Mineralisation occurs in the Cahill Formation metamorphic rocks just above a reverse faulted contact with the Kombolgie Formation sandstone. Primary mineralisation as 1% uranite ($U_3O_8$) occurs as an elongated wedge approximately 100 m long and up to 50 m wide.  It has a dip of 55° and tapers out at about 100 m depth.  It is confined to a series of highly chloritised quartz schists.  The ore body is weathered to a depth of about 25 m to form a secondary ore zone which overlies the primary zone and extends 100 m as a horizontal tongue in weathered rock, down slope and south east of the primary zone.

The groundwater hydrology is quite complex with sources and sinks being dependent on the season.  There are two aquifers, a superficial aquifer which is present only during the wet season and there is a permanent Cahill aquifer at depth.  The Cahill aquifer is confined by the overlaying weathered zone which acts as a leaky aquitard.  The superficial aquifer is associated with the surface deposits of the area, these deposits have a high infiltration capacity and permeability.

At the end of the dry season the surface deposits dry out and the water table sinks by up to 9 m.  With the onset of the wet season the surface deposits receive rain water and runoff from the low permeability Kombolgie sandstone.  The water table rises and the surface deposits saturate and, where the water table progressively cuts the surface, release water, which drains into the local creek system in a south easterly direction. With the onset of the dry season the surface deposits dry out by drainage and evaporation.

The Cahill aquifer is associated with the unweathered quartz schists of the Cahill Formation. This formation has hydraulic continuity and an average bulk transmissivity of 20 m²/d, with higher values in fracture zones. Flow is in a south easterly direction. It is recharged by the Cahill aquifer up gradient of the ore body, by infiltration from the surface deposits adjacent to the Koongarra Reverse Fault and by the Kombolgie Formation at depth. Although some water drains into the local creeks through leakage into the surface deposits, most of it is believed to drain into subsurface dolomites south east of the site.

Four bores running down gradient from the centre of the ore body were selected. Unfortunately, the furthest in-line bore could not be located so bore PH139 was substituted. The bores which had been sunk by percussion drilling, were around 80 m in depth and cased to about 30 m depth which was the limit of the weathered zone. Pump tests indicated that recharge was principally from the first 10 m below the casing. It may be assumed that the sample represents the top of the Cahill aquifer together with a fraction of the superficial aquifer.

**Solute Chemistry**

The ground water solute chemistry for the two sites is given in Table 2. The waters were of low salinity, with a pH range of 6.5 to 7.5, and slightly more acidic closer to the ore bodies, reflecting the acidic nature of the deposits. In general, the Eh indicated that the waters were oxidising, although at Koongarra a fraction of the water was derived from the deeper, reduced aquifer.

**Colloid Chemistry**

The colloids were characterized by analysing the ultrafiltrates and colloid concentrates for elemental and isotopic differences. Elemental analysis included Fe, Al and Si as hydrolysing oxides and Ca, Mg, Na and K as components of colloidal clay. A plot of ultrafiltrate concentration versus colloid concentration is given in Figure 5. The Al levels in all the samples were below the point of detection by atomic absorption. The results show that the only element significantly enriched in the colloid fraction was iron.

There was a marginal enrichment of Si but the results were very close to the limit of detection.

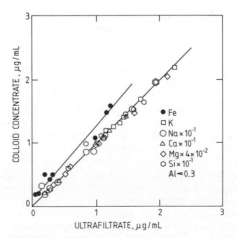

Figure 5.    The element distribution between colloid concentrate and ultrafiltrate.

Table 2.    Solution Parameters and Chemical Analysis (mg/L) of Ground Waters

| Bore | pH | Eh (mV) | Conductivity (μs) | Ca$^{2+}$ | Mg$^{2+}$ | Na$^+$ | K$^+$ | HCO$_3^-$ | Cl$^-$ | SO$_4^{2-}$ |
|------|----|---------|-------------------|-----------|-----------|--------|-------|-----------|--------|-------------|
| **NABARLEK** | | | | | | | | | | |
| OB19 | 6.6 | 335 | 145 | 4 | 86 | 16 | 1 | 340 | 116 | 1 |
| OB25 | 6.8 | 315 | 175 | 15 | 62 | 11 | 2 | 265 | 84 | 2 |
| RN21038 | 6.9 | 335 | 145 | 11 | 39 | 10 | 1 | 220 | 25 | 1 |
| RN20475 | 7.2 | 305 | 165 | 12 | 41 | 9 | 1 | 270 | 14 | 2 |
| **KOONGARRA** | | | | | | | | | | |
| PH49 | 6.5 | 170 | 105 | 2 | 24 | 1 | 1 | 54 | 14 | 10 |
| PH55 | 6.8 | 375 | 74 | 2 | 21 | 2 | 1 | * | 7 | 9 |
| PH69 | 6.9 | 250 | 80 | 2 | 21 | 3 | 1 | 71 | 5 | 3 |
| PH139 | 7.2 | 370 | 88 | 2 | 14 | 4 | 1 | 56 | 4 | 1 |

## Isotope Chemistry

**Uranium solute**. The uranium isotope results are listed in Table 3. At both sites, the solute uranium concentration decreased logarithmically with distance to background levels down the hydraulic gradient. The decrease is typical of a dispersion fan. The $^{234}U/^{238}U$ activity ratio of the solute shows a slight rise with distance away from the ore body. This is a common occurrence and is well documented in the literature (Osmond and Cowart, 1976). Because of the limited number of results, no correlations are reported for the $^{234}U/^{238}U$ ratio versus the uranium concentration or its inverse.

**Uranium colloid**. There was a slight uranium enrichment in the colloid concentrate, ranging from 2 per cent at Nabarlek up to 11 per cent at Koongarra. There was no correlation between uranium transported as colloid and the solute concentration.

The $^{234}U/^{238}U$ activity ratio was usually less than or in the vicinity of 1. It was independent of distance from the ore body and, of the uranium concentration either in the solute or in the colloid. The activity ratios for the colloid for the near field samples were generally lower than those for the solute. However, the differences in values are usually within the range of the propergated errors and so cannot be considered significant. For the far field samples, and for a very limited number of results, the solute and colloid $^{234}U/^{238}U$ activity ratios started to diverge indicating that there was an increasing independence between solute and colloid transport.

**Thorium solute**. Tables 4 and 5 list the results for $^{232}Th$ and $^{230}Th$, respectively. The $^{232}Th$ solute concentration is independent of the presence of other thorium isotopes and distance down gradient of the ore bodies. The thermodynamic solubility limit for Th as thorianite is $1x10^{-11}$ g/L which is equivalent to $2.4x10^{-6}$ (dpm/L) $^{232}Th$ (Langmuir and Herman, 1980). The analyses indicate that the $^{232}Th$ concentration is about three orders of magnitude above the thermodynamic solubility limit, probably because of complexation by trace organics or phosphates. The values are typical of those reported in the literature for environmental levels of thorium in fresh waters.

The $^{230}Th$ solute concentration decreased logarithmically with distance down gradient of the ore body at both sites. The solute $^{230}Th$ was in gross activity deficiency to that of its parent $^{234}U$, and the $^{230}Th/^{234}U$ activity ratio increased down gradient of the ore body. The solute $^{230}Th$ concentration was at least one order and sometimes up to two orders of magnitude below the thermodynamic solubility limit of 0.431 dpm/L for $^{230}Th$. It is, therefore, necessary to account for the very low level of $^{230}Th$ in solution when there is an active source of $^{230}Th$ in the form of ingrowth from $^{234}U$. The solubility of the ingrowing $^{230}Th$ is apparently controlled by the preexisting complexed $^{232}Th$. If the

Table 3. Uranium-238 as Solute and Colloid in Ground Water

| Bore Hole | 238U as Solute | | 238U as Colloid | | % 238U in colloid form |
| | dpm/L | AR | dpm/L | AR | |
|---|---|---|---|---|---|
| **NABARLEK** | | | | | |
| OB19 | 52.5 ± 2.5 | 1.051 ± 0.071 | 0.238 ± 0.087 | 0.85 ± 0.49 | 0.45 ± 0.17 |
| OB25 | 6.92 ± 0.25 | 1.081 ± 0.054 | 0.136 ± 0.006 | 0.788 ± 0.060 | 1.93 ± 0.11 |
| RN21038 | 2.088 ± 0.088 | 1.032 ± 0.062 | 0.0088 ± 0.0030 | 0.90 ± 0.46 | 0.42 ± 0.14 |
| RN20475 | 0.0814 ± 0.0062 | 2.04 ± 0.19 | 0.0017 ± 0.0002 | 0.17 ± 0.17 | 2.05 ± 0.28 |
| **KOONGARRA** | | | | | |
| PH49 | 149.3 ± 8.1 | 0.999 ± 0.077 | 1.95 ± 0.15 | 0.98 ± 0.11 | 1.29 ± 0.12 |
| PH55 | 0.325 ± 0.013 | 1.011 ± 0.059 | 0.0198 ± 0.0012 | 0.984 ± 0.083 | 5.74 ± 0.12 |
| PH58 | 0.217 ± 0.009 | 0.989 ± 0.061 | 0.0272 ± 0.0006 | 1.13 ± 0.03 | 11.1 ± 0.5 |
| PH139 | 0.0368 ± 0.0046 | 1.29 ± 0.21 | 0.00115 ± 0.00009 | 1.04 ± 0.12 | 3.03 ± 0.44 |

Note:
(1) All ± errors quoted are 1 fully propagated to the final result.
(2) dpm = disintegrations per minute
(3) AR = Activity Ratio 234U/238U

Table 4.    Thorium-232 as Solute and Colloid in Ground Water

| Bore Hole | $^{232}$Th as Solute dpm/L | $^{232}$Th as Colloid dpm/L | % of total |
|---|---|---|---|
| **NABARLEK** | | | |
| OB19 | 0.0018 (99%C) | 0.00013 $\pm$ 0.00003 | 6.8 (99%C) |
| OB25 | 0.0020 $\pm$ 0.0012 | 0.00005 (99%C) | 2.5 (99%C) |
| RN21038 | 0.0019 $\pm$ 0.0008 | 0.000026 $\pm$ 0.000024 | 1.4 $\pm$ 1.4 |
| RN20475 | 0.0009 $\pm$ 0.007 | 0.0000025 (99%C) | 2.8 (99%C) |
| | | | |
| **KOONGARRA** | | | |
| PH49 | 0.0015 $\pm$ 0.0011 | 0.00003 $\pm$ 0.00002 | 2 $\pm$ 2 |
| PH55 | 0.0040 $\pm$ 0.0006 | 0.00011 $\pm$ 0.00002 | 2.6 $\pm$ 1.6 |
| PH58 | 0.0005 $\pm$ 0.0005 | 0.00011 $\pm$ 0.00002 | 18 $\pm$ 15 |
| PH139 | 0.0009 $\pm$ 0.0005 | 0.00002 $\pm$ 0.00001 | 2.2 $\pm$ 1.6 |

Note:  (99%C) signifies value is below the 99% confidence limit given.

kinetics of isotopic exchange between complexed $^{232}$Th and noncomplexed $^{230}$Th is slower than the kinetics for the adsorption of $^{230}$Th onto surfaces of colloids and pore walls, then there would be preferential loss of $^{230}$Th from solution onto local surfaces. This would account for the enhanced $^{230}$Th observed by Lowson et al. (1986) in iron phases down gradient of the neighbouring Ranger ore body.

**Thorium colloid**. The concentration of $^{230}$Th carried as colloid is listed in Table 5.  The $^{230}$Th colloid concentration decreases with distance, although there are some large fluctuations.  Up to 47% of the total $^{230}$Th was observed to be transported in the colloidal form.  The percentage as colloid was independent of distance or $^{230}$Th solute concentration.  The $^{230}$Th/$^{234}$U activity ratio in the colloid phase increases with distance down gradient of the ore body.  The colloid has a significantly higher $^{230}$Th/$^{234}$U ratio than the solute.  This indicates that the $^{230}$Th in the colloid is independent of the $^{230}$Th in the solute and supports the suggestion from the uranium results that the colloid should be considered as a separate phase, with minimal interaction with the solute, and that the $^{230}$Th is ingrown from uranium entrained in the colloid.

**Nature of the colloid**. Table 6 lists the molar concentrations of the enhanced elements in the colloid concentrate.  The results show that the dominant element is iron, although silicon (which was at the detection limit) may be present in similar proportions. Uranium and thorium are only present as trace quantities.  From this, it is inferred that the colloid is a microcrystalline form of ferric hydroxide, commonly described as ferrihydrite (Chukhrov et al., 1972) in association with polymeric forms of silica or silicic acids.

## CONCLUSIONS

Ground water colloids are present at very dilute concentrations.  This makes detection very difficult.

Combining radiochemical analysis with large sample ultrafiltration has proved to be a useful method of ground water colloid characterization.  The techniques have allowed the following conclusions to be drawn.

1.    A colloidal component within the range 1 μm to 18 nm has been identified by the ultrafiltration of ground waters sampled down gradient of two Australian uranium ore bodies.

Table 5. Thorium-230 as Solute and Colloid in Ground Water

| Bore Hole | 230Th as Solute | | 230Th as Colloid | | % 230Th in colloid form |
|---|---|---|---|---|---|
| | dpm/L | AR | dpm/L | AR | |
| **NABARLEK** | | | | | |
| OB19 | 0.0255 ± 0.0026 | 0.00046 ± 0.00005 | 0.0064 ± 0.0002 | 0.028 ± 0.013 | 20.3 ± 1.8 |
| OB25 | 0.0095 ± 0.0031 | 0.0013 ± 0.0004 | 0.00066 ± 0.00010 | 0.0061 ± 0.0010 | 6.4 ± 2.2 |
| RN21038 | 0.0079 ± 0.0020 | 0.0037 ± 0.0009 | 0.00122 ± 0.00010 | 0.154 ± 0.060 | 13.2 ± 5.3 |
| RN20475 | 0.0012 (99%C) | 0.0069 (99%) | 0.00064 ± 0.00007 | 0.521 ± 0.124 | 36 (99%) |
| **KOONGARRA** | | | | | |
| PH49 | 0.0123 ± 0.0028 | 0.00008 ± 0.00002 | 0.0111 ± 0.0004 | 0.0058 ± 0.0018 | 47 ± 6 |
| PH55 | 0.0230 ± 0.0035 | 0.070 ± 0.011 | 0.0039 ± 0.0002 | 0.200 ± 0.015 | 14.5 ± 2.0 |
| PH58 | 0.0131 ± 0.0025 | 0.061 ± 0.012 | 0.0081 ± 0.0003 | 0.265 ± 0.0112 | 38.2 ± 4.7 |
| PH139 | 0.0045 (99%C) | 0.096 (99%C) | 0.00142 ± 0.0000000008 | 1.19 ± 0.11 | 24 (99%C) |

Note:
(1) All ± errors quoted are 1 fully propagated to the final result.
(2) dpm = disintegrations per minute
(3) AR = Activity Ratio 234U/238U
(4) (99%C) signifies value is below 99% confidence limit given

Table 6.    Molar Concentrations of Enhanced Elements in the Colloid Phase (Mole/L)

| Site | Fe (10-9 M) | 238U (10-9 M) | 232Th (10-15 M) |
|---|---|---|---|
| **NABARLEK** | | | |
| OB19 | 51.5 | 1.3 | 223 |
| OB25 | 38.7 | 0.77 | 8.8 |
| RN21038 | 50.1 | 0.049 | 4.5 |
| RN20475 | 38.7 | 0.0095 | 0.46 |
| | | | |
| **KOONGARRA** | | | |
| PH49 | 38.3 | 11.01 | 5.3 |
| PH55 | 141 | 0.11 | 19 |
| PH58 | 87.0 | 1.53 | 10 |
| PH139 | 60.9 | 0.0065 | 3.9 |

2.    The colloid is composed largely of iron and silicon species with entrained uranium and thorium.

3.    Only a minor proportion of the total uranium is associated with the colloid but a significant proportion of the $^{230}Th$ is associated with the colloid.

4.    Solute $^{232}Th$ is present at up to two orders of magnitude above the solubility limit for $ThO_2$. This is attributed to complexation by trace organics and phosphates.

5.    Solute $^{230}Th$ is present at one to three orders of magnitude below the solubility limit for $ThO_2$ and is in large deficit to its solute parent $^{234}U$.

6.    The high loss of $^{230}Th$ from solution is due to fast adsorption kinetics compared to slow isotopic exchange kinetics with complexed $^{232}Th$.

**ACKNOWLEDGEMENTS**

This project was funded by the United States Nuclear Regulatory Commission as part of Contract No. NRC-04-81-172.

**REFERENCES**

Benes, P. and Majer, V., 1980. "Trace Chemistry of Aqueous Systems", Elsevier, Amsterdam.

Cooper, A.R., 1984. Ultrafiltration. *Chem. Brit.*, 20(9):822-830.

Chukhrov, F.V., Zvagin, B.B., Ermilova, L.P. and Gorshkov, A.I., 1972. *in:* "Proc. Internat. Clay Conf.", Madrid, pp 333-341.

Fane, A.G., Fell, C.J.D. and Waters, A.G., 1981. The relationship between membrane surface pore characteristics and flux for ultrafiltration membranes, *J. Membr. Sci.*, 9:245-262.

Kepak, F., 1971. Adsorption and colloidal properties of radioactive elements in trace concentrations. *Chem. Rev.*, 71:357-369.

Langmuir, D. and Herman, J.S., 1980. The mobility of thorium in natural waters at low temperatures, *Geochim. Cosmochim. Acta*, 44:1753-1766.

Liesner, K.H., Gleitsmann, B., Peschke, S. and Steinkopff, Th., 1986. Colloid formation and sorption of radionuclides in natural systems, *Radiochemica Acta*, 40:39-47.

Lowson, R.T., Short, S.A., Davey, B.G. and Gray, D., 1986. $^{230}U/^{234}U$ and $^{230}Th/^{234}U$ activity ratios in mineral phases of a lateritic weathered zone, *Geochim. Cosmochim. Acta*, 50:1696-1702.

Osmond, J.K. and Cowart, J.B., 1976. The theory and uses of natural uranium isotopic variations in hydrology. *Atomic Energy Rev.*, 144:621-679.

Paneth, F., 1913a, A new method for concentrating polonium, *Monatsh.*, 34:401.

Paneth, F., 1913b, Colloidal solutions of radioactive substances, *Kolloid Z*, 13:297.

Salbu, B., Steines, E. and Bjornstad, H., 1984. Use of different physical separation techniques for trace element spectation studies in natural waters, *in:* "Hydrochemical Balances of Fresh Water Systems. Proc. Uppsala Symposium", IAHS-AISH Publication Vol. 150: Uppsala, pp 203-213.

Yariv, S and Cross, H., 1979. "Geochemistry of Colloid Systems", Springer-Verlag, Berlin.

# SECTION II


# WATER TREATMENT PROCESSES

# WATER TREATMENT TECHNOLOGY IN AUSTRALIA

Brian A. Bolto

CSIRO Division of Chemicals and Polymers
Clayton, Victoria

## INTRODUCTION

Australia is the driest continent in that it receives a small and variable rainfall (only Antarctica has less), and there is poorer surface run off than for any other continent. Some 89% of precipitation is lost by evaporation and infiltration. Because of limited dilution and the generally low flows, streams and water bodies are especially vulnerable to pollution (Pigram, 1985). The variability also makes for peaks and troughs in the pollution load. As well as watercourses being far more variable than in other countries, the variability here increases as catchment size increases, which is the opposite of the usual pattern (Peters, 1986).

Australian water resources have been the subject of many reviews over the past decade. The most relevant to the present topic are those on water quality and water technology in the *Water 2000* series of reports (Garman et al., 1983; Gutteridge et al., 1983).

### Contaminants Present

Naturally occurring waters can be polluted with a great variety of materials (Pigram, 1985; Gutteridge et al., 1983). After removal of gross contaminants such as animals and detritus there may be suspended and dissolved impurities, including

- turbidity particles, which protect microbes from disinfectants (e.g. clay, hydrous metal oxides)

- microbes such as viruses, bacteria, amoeba, algae

- coloured compounds, both colloidal and dissolved, which form trihalomethanes (THMs) on chlorination

- organic compounds causing gastric upsets in humans, stock kills and taste and odour

- inorganic chemicals such as salts, hardness, metal ions, alkalinity, nitrate, fluoride, radioactive species

All have public health significance, except iron, manganese and taste and odour compounds, which can be nuisance materials for aesthetic reasons.

*Surface and Colloid Chemistry in Natural Waters and Water Treatment*
Edited by R. Beckett, Plenum Press, New York, 1990

Table 1.   Australian Guidelines for Drinking Water Quality

| | |
|---|---|
| turbidity | 5 NTU (nephelometric turbidity units) |
| coliforms | 10 in 100 mL |
| colour | 15 PCU (filtered; platinum cobalt units) |
| THMs | 200 µg/L |
| TDS | 1000 mg/L |
| Sodium | 300 mg/L |
| hardness | 500 mg/L (as $CaCO_3$) |
| nitrate | 10 mg/L (as N) |
| fluoride | 0.5 to 1.7 mg/L |

Many pollutants arise from human activities.  These include, in addition to some of the above,

- nutrients from fertilizers and sewage effluents i.e. nitrogen and phosphorus compounds

- carbonaceous impurities from sewage effluents, intensive livestock enterprises and food processing

- toxic chemicals from agricultural areas, mining sites and industry

- suspended matter and salinity from land clearing and over-irrigation

The problems of wastewater treatment will not be addressed here in any detail; the emphasis will be on purification of low grade waters for water supply purposes.

**Water Quality Issues**

Australia's water quality problems are myriad (Pigram, 1985; Garman et al., 1983).  Contamination of supplies with turbidity particles is widespread in towns which use river waters as their water source, notably in the Murray-Darling Basin. Microbiological hazards with such turbid waters abound in Victorian country areas, where the majority of towns fail to meet quality targets for most of the time (Curtin and McNeill, 1987).  Colour is a ubiquitous difficulty, but a particular one in Tasmania; salinity has its worst effects in the Murray-Darling Basin and Western Australia, but is present in most of inland Australia.

The new Australian guidelines for desirable drinking water quality (National Health and Medical Research Council, 1987) are listed in Table 1 for major contaminants:

They appear to be liberal on turbidity, but for the major urban areas a level of less than 1 NTU is preferred so that disinfection is secure and complete.  The THM level is more than double European and North American limits, and the TDS level has been set at 1000 instead of 500 mg/L, which accepts the real situation presently existing.

Current surveys of water treatment practices across the nation being conducted by the Australian Water Resources Council (AWRC) on a State-by-State basis show that 0.5 to 6% of the population receive water of unsatisfactory quality (Garman, 1987), which is probably about a third of the rural population.  It is estimated that nationwide at least 250,000 people receive an untreated supply (Pigram, 1985).  The anticipated cost to upgrade inferior supplies is at least $2 billion, as reported in the *Water 2000* studies.

**Water Purification**

In water purification it is preferable to remove the small amounts of offending impurities rather than extract or separate the large volume of water.  Reverse osmosis

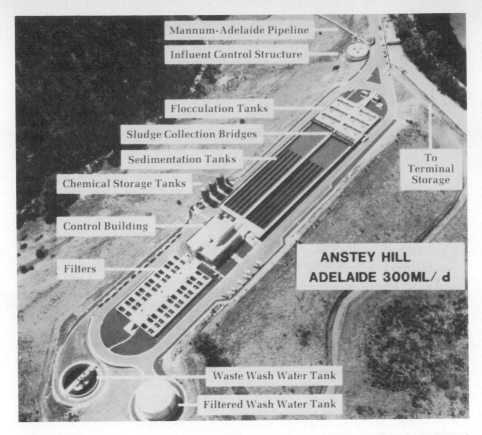

Figure 1.    Plan view of the Anstey Hill water treatment plant (Engineering and Water Supply Department of SA, 1980).

and distillation methods of desalination are exceptions to this general rule.  Costs can be kept down by using simple and compact equipment, keeping energy and chemical usage to a minimum and if possible recycling the reagent employed, and avoiding waste disposal problems posed by sludges and large volumes of effluents.

The general aim of water purification cannot be better expressed than in the first known patent on the topic (Baker, 1981), granted to William Walcott in England in 1675:

*"..... on the art of making corrupted water fit for use, and .... fresh, clean and wholesome in very large quantities, by such wayes and means as are very cheap and easy, and which may be done and practiced with great speed....."*

## REMOVAL OF TURBIDITY, MICROBES AND COLOUR

In this important area of water purification, the conventional method involves adding a coagulating agent such as an aluminium or iron salt to the raw water at an appropriate pH level.  A metal hydroxide floc is precipitated which has a positive surface charge.  The turbidity particles, microbes and colour bodies all have a negative surface charge in normal pH regimes, so they adhere to the floc.  Coagulation and flocculation proceed, followed by sedimentation and filtration through sand to remove any carryover floc.  A neutral polymer or anionic polyelectrolyte in addition to aluminium or iron salts may be beneficial.  Activated silica is an alternative secondary coagulant.

## Conventional Systems

Very large scale water treatment plants make use of rectangular sedimentation tanks which have a horizontal water flow. For example, that at Anstey Hill in Adelaide treats 300 ML/d and has twin tanks 90 m long, 17 m wide and 5 m deep. Figure 1 shows the layout, with collection troughs visible over most of the sedimentation tanks (Engineering and Water Supply Department of SA, 1980). Filters at the bottom end of the plant are followed by a waste wash water tank which is for storage of water from the backwashing of the sand filters; the backwashing is carried out on average every 40 h. Typical chemical doses are 45 mg/L of alum and 2 mg/L of activated silica, to treat a raw water from either the Murray River or from reservoirs, with the turbidity varying from 6 to 200 NTU, and the colour 15 to 100 PCU. The product water has turbidity 0.5 NTU and colour 10 PCU. The plant cost $20 million in 1980.

Sedimentation systems can also be circular in plan, with radial collection troughs (James M. Montgomery Inc., 1985); they still require a large land area, but are commonly used for installations of throughput smaller than 100 ML/d. In these circular clarifiers there can be a combination system which has a rapid mixing compartment at the centre, with a slow mixing and floc formation compartment surrounding it. Water flow is then upwards from the base of the tank through a sludge blanket which comprises a solids contact system. The water flow is vertical in this arrangement.

Clarifiers generally remove 90% of the impurity-loaded floc. Sand filters remove the remainder; they are often dual media systems, made up of an anthracite layer of 50 cm depth over a sand layer of 25 cm, and run at a high rate of 10-15 m/h (James M. Montgomery Inc., 1985). Polymer dosing ensures that the floc does not break through the filter. Batch operation is utilized, with filter runs of one to two days, when backwashing is necessary to clear the filter bed.

## Dissolved Air Flotation

When algae are present in the raw water a flotation method may be preferred, instead of settling flocs by gravity. Algae often do not settle sufficiently because of attachment to oxygen bubbles, or they may block filters too rapidly (Williams et al., 1985). In dissolved air flotation, air is dissolved under pressure in water which is then discharged into the floc-carrying water. Microbubbles come out of solution and buoy the flocs and algae to the surface where they are removed with scrapers. About 10% of the product water is recycled for air dissolution, which makes for high energy costs in what is a low capital cost process.

## Direct Filtration

For waters of moderate turbidity and colour the flocculation tank/clarifier may be omitted. The raw water turbidity level should be less than 15 NTU and the colour less than 40 PCU (Gutteridge et al., 1983). A low alum dose is employed and after in-line mixing and a brief period (15-20 min) in a contact tank the dosed water goes to sand or dual media filters. To strengthen the microflocs produced a nonionic polymer or activated silica is usually added. The microflocs are retained in the filter bed by an adhesion mechanism; the filter media are designed to trap the fine floc and store it. Regular backwashing is required, at intervals of 12 h to several days. There are considerable capital cost savings of about 30%. A number of direct filtration plants are in operation in Australia (Strom et al., 1985). However, it must be stressed that the raw water should be relatively free of algae and any suspended matter (Huber, 1985), including microbial cells and precipitated manganese dioxide.

## Slow Sand Filtration

Of course one may not add chemicals at all, omit the coagulation step, and merely perform the filtration step. Slow sand filtration was first used at the start of the last century. A 1 m deep bed of sand is employed, over a layer of gravel. Slimes develop on the sand which aid the process, and protozoa present consume bacteria. It is generally

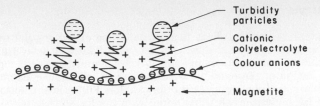

Figure 2.  Diagrammatic representation of colour and turbidity removal by magnetite.

necessary to clean the bed by scraping off the top layer of sand every 6 to 12 weeks. The experience from five Australian plants is that filter runs are as short as three days if high turbidity peaks occur in the raw water (Australian Water Resources Council, 1987). Algae can be removed first from the raw water by dissolved air flotation, as has been shown recently at Brisbane where a slow sand filtration plant has been in operation for some 70 years. Another approach is to add a roughing prefilter, and a unit of this type has been tested at Pyalong, Victoria (Drew et al., 1987). The horizontal prefilter consists of three compartments containing successively diminishing sizes of gravel. A raw water turbidity of 8 NTU is reduced to 2 NTU, and colour from 25 to 10 PCV by the integrated system. The flow rates are 2 m/h for the roughing and 0.1 m/h for the slow sand filter.

## New Approaches

In the fully fledged alum process used to obtain a high quality water as described in the Section above on conventional systems, sedimentation is slow, and a voluminous sludge containing the highly swollen aluminium hydroxide causes a disposal problem. A large land area is required for the plant, and filtration is essential as settling of the floc is never perfect. To overcome these disadvantages CSIRO studies on magnetic reagents for water treatment were extended to fine, rapidly reacting magnetite ($Fe_3O_4$) particles. The outcome is a process based on a rapidly settling and recyclable reagent (Kolarik et al., 1977).

## CSIRO research

The CSIRO research on new techniques for water purification in general is based on insoluble reagents such as metal oxides or crosslinked polymers (Bolto, 1983). The reagents are reusable, so chemical usage is minimized, which keeps costs down. More importantly, micro-sized particles are employed, which means that reactions are fast and the consequent increase in the intensity of processing means that equipment needs and hence costs are lower. Because such particles are slow settling or require high pressures in packed column usage, unconventional methods are needed to handle the microparticles.

Using a magnetic particle as a coagulant or incorporating a magnetic filler in an adsorbent offer new approaches to separation problems. The concept has been employed in a number of physico-chemical processes in two broad categories: the removal of insoluble impurities by coagulation and the uptake of dissolved contaminants by adsorption.

## The SIROFLOC process

Magnetite particles of size 1-10 μm are employed at pH 5-6. A 1% slurry in the raw water will remove about half the turbidity and colour because of the positive surface charge on the magnetite. In a typical example the addition of 1 mg/L of a cationic polyelectrolyte will result in the remaining impurities being heterocoagulated with the magnetite, to give a product water of turbidity 1 NTU and colour 4 PCU from an original raw water of turbidity 34 NTU and colour 43 PCU (Anderson et al., 1981).

The model for the process in terms of charge interactions is shown in Figure 2.

The colour anions and turbidity particles are negatively charged and adhere to the positively charged magnetite (only the anions are depicted for simplicity). As a result the charge on the loaded magnetite surface becomes negative. The addition of a cationic polymer binds the remaining impurities by bridging the two negatively charged entities.

The dense nature of the loaded magnetite makes for rapid separation from the purified water; settling occurs at three times the rate of an alum floc, which is increased to six times when the magnetite is magnetized. This is because a magnetic network of particles is produced, as depicted in Figure 3. In practice, a simple permanent magnet around the exit pipe from the contact tanks achieves a suitable effect.

The loaded magnetite is treated with alkali at pH 11-12 which causes the surface charge on the iron oxide to become negative, whereupon the other species (whose charge remains unaltered) are sloughed off from the surface. After washing of the alkali-treated magnetite on magnetic drum separators it is returned by pumping, which demagnetizes it, to the start of the process. The colloid science of the process is discussed in more detail in the Dixon and Kolarik paper in this volume.

Two large-scale plants are now in operation, a 35 ML/day installation at Mirrabooka in Western Australia, and a 20 ML/day unit at Bell Bay, Tasmania (Nadebaum and Fish, 1985). They are unique because of the

- rapid sedimentation, so the clarifier is less than half the normal size

- absence of a sludge which requires disposal

- omission of the filtration step

- small land area required

- reuse of the primary coagulant, magnetite

- quick start up, so that 8 h/day operation is feasible if it is desired.

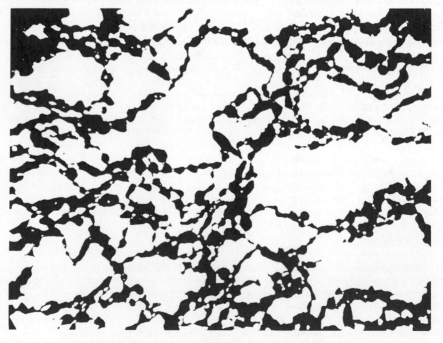

Figure 3.    Flocculation of magnetized particles.

Depending on circumstances, the capital cost may be lower than that of conventional treatment by 30%. A further plant of 20 ML/d size is due for completion at Redmires in Yorkshire in 1988, following endorsement of the process by the Water Research Centre in the UK. The process is an answer to the current high levels of aluminium and heavy metals which are of concern in Europe and North America.

## Comparison of Water Treatment Capital Costs

With full conventional treatment, comprising the three stages of coagulation/flocculation, sedimentation and filtration, plant capital costs are highest for the very large-scale installations of greater than 100 ML/d capacity, where huge rectangular sedimentation tanks are employed. For small units which can make use of circular clarifiers there is a relative cost saving advantage (Gutteridge et al., 1983).

By far the greatest economies, amounting to about a 30% cost reduction, are achieved by omitting one of the three stages. Examples are direct filtration, if it is permitted by the raw water quality (no sedimentation stage) and SIROFLOC regardless of raw water quality (smaller sedimentation system and no filtration stage). As an interim measure, for small plants of 5-10 ML/d capacity the installation of filters may be postponed if a lower product standard can be tolerated. A product water level to 3 NTU turbidity is claimed when there is no filtration (Gutteridge et al., 1983).

## Iron and Manganese Removal

The oxidation of soluble reduced forms of iron and manganese to insoluble hydrous oxides is catalysed by bacteria, which can grow prolifically in the reticulation system. They then present aesthetic problems of taste and laundry staining.

For surface waters, natural oxidation can be encouraged by destratifying reservoirs. If necessary, chemical oxidation can be employed at the treatment plant. Chlorine or air may be used to oxidize iron. In the case of air the oxidation occurs after precipitation of the hydroxide with lime. Manganese requires stronger oxidants such as permanganate, ozone, or chlorine dioxide. For the latter contaminant, the precipitated manganese dioxide catalyses further oxidation of manganous ions. After clarification it is essential to rigorously disinfect the product water to minimize bacterial growth in pipelines. The taste thresholds for iron and manganese are 0.3 and 0.1 mg/L respectively. Manganese is a particular problem in Queensland (Loos, 1987).

## Country Town Supplies

A lot of thought has gone into the provision of water supplies of adequate quality to country towns, especially those that have no treatment at the moment. A quality target inferior to that for major urban centres is possible, such as turbidity 2 NTU instead of 1 NTU. Staging of treatment by omitting the filter is a temporary expedient, but will not meet a target of 2 NTU. The SIROFLOC method loses its economic advantage on the small scale. Suitable technologies are slow sand filtration, but with a pretreatment probably necessary for most Australian conditions, especially when eutrophic waters are involved. Dissolved air flotation or a roughing prefilter are then advocated. Dual reticulation is appropriate to conserve the high quality treated water for the kitchen and laundry, provided there is not an extended pipeline system. The low quality raw water is used for toilet and garden.

Other possibilities are electrocoagulation and continuous contact filtration. Point-of-use treatment will be dealt with in a later section of the paper.

**Continuous contact filtration.** A continuous version of direct filtration eliminates backwash storage, and all operations are carried out in the one vessel (Larsson, 1987). There are no coagulation/flocculation or sedimentation tanks. High turbidity waters can be treated with 20-30% less coagulant than is needed for full conventional treatment. This Swedish technology is worthy of investigation for water treatment in Australia; two such plants have recently been installed. The method has been utilized for treating mining effluents.

Coagulation, flocculation and separation all take place in the filter bed. There is upflow of the dosed water through the descending sand bed, with the dirtiest sand at the bottom being withdrawn and cleaned by a turbulent flow of air and pushed up to a sand washer. After washing it flows down to the top of the bed. Product and wash water are taken off continuously at the top of the filter column.

**Electrocoagulation.** Galvanic action in a mixture of aluminium and coal particles will produce aluminium ions and hence an aluminium floc, while oxygen is reduced to water at the coal surface, so that the pH level rises (Collins and Johnson, 1985). Clarification can be achieved in a filter bed made up of the mixed particles if there is a low flow rate or in a separate vessel/filter at high flow rates. The approach is under further study at CSIRO (Kolarik and Chin, 1987).

## Disinfection

Of the range of chemicals available, chlorine is the most effective in terms of minimum THM formation, residual protection and cost efficiency, *provided* that adequate colour removal has been achieved beforehand. All methods require a low turbidity in the water for reasonable dose levels and efficient kills of microbes.

Chloramination is an age-old process which fell into decline but is now being resurrected. It is not as simple operationally as chlorination, but has high persistence so it is relevant for long pipeline systems, and also for poor quality waters which give high THMs with chlorine. Concurrent addition of ammonia and chlorine is the best approach, but there are a number of problems, including odours, nitrification, corrosion, fish kills and requirements for kidney dialysis users.

Chlorine dioxide gives better virus and bacteria kills than the preceding reagents, and does not intensify odour problems. It does not react with ammonia, nitrogen compounds or phenols. However, it is costlier than most other reagents, being made from chlorine and sodium chlorite.

Ultraviolet light definitely demands a low turbidity water and is not often used on the large scale. There is no residual, so chlorine must be added afterwards.

Ozone is very effective and removes taste and odour compounds as well, but it results in a more complex process, is costly to use and leaves no residual, which must be provided separately with chlorine. Preozonation enhances flocculation, and its use for this purpose is expanding (Glaze, 1987).

## REMOVAL OF SALTS, HARDNESS, TASTE AND ODOUR

The removal of dissolved contaminants requires quite different procedures. The needs are not as pressing in Australia as they are for the removal of insoluble impurities, which are generally a more immediate health hazard. Included in this category are alkalinity, heavy metal ions, radioactive species, nitrate and fluoride.

## Desalination

Australia has a long desalting history, going back to the turn of the century. In the goldfields of Western Australia distillation of saline ground water was practised, but the efforts were short lived because of the scarcity of timber for use as fuel. Long pipelines from dams near the coast were the solution of the day.

Salinity problems exist in all States, with the most marked being in the Murray-Darling Basin, Western Victoria and the south west of Western Australia. There are about 60 desalination plants of capacity 10 kL/d or greater, which, although only 1% of total world plant numbers, has doubled in the last decade (Bolto, 1984). Most units are installed at sites of mineral developments and tourist resorts, and at off-shore oil platforms.

The predominant techniques are membrane based, especially those for the treatment of brackish waters. Here reverse osmosis is the most used (50 units approximately), although electrodialysis is a recent contender. Ion exchange is applicable to dilute waters, but chemical costs make its use uncompetitive. Distillation is only used where there is readily available fuel, as at gas fields, and is utilized mainly for sea water. Reverse osmosis for sea water is a recent innovation. Desalination is an energy intensive process, whatever the type of energy required, and is the most expensive unit process in water treatment.

**Reverse osmosis.** In reverse osmosis, sufficient pressure must be applied to the saline water to force water through a thin nonporous membrane at a suitable rate (Crossley, 1983). Three types of membrane materials are used

- cellulose acetate, which is pH sensitive and susceptible to bacterial attack

- polyamide, which is chlorine sensitive and subject to biofouling

- thin film composites, such as polyamide on porous polysulphone

The pressures needed are 2-8 MPa for brackish water, and 7 MPa for sea water. Energy recovery systems based on turbines have been developed, but they add to the capital costs. Lower pressure membranes are being developed too, which require 40% less energy; the recovery of product water is lower though, at 75% instead of 85% in one example (Crowdus, 1984).

Spiral wound formats are the most common (Crossley, 1983). The feed water flow is through the end of the spiral roll, along a spacer between the supported membranes, through the membranes on either side of the support, and then via the porous support to an exit tube in the centre of the roll.

To minimize scaling and fouling, pretreatment of the feed water is essential. Turbidity and colour should be reduced. Calcium and silica levels are critical, so it is desirable to soften the water or add antiscalant polymers. The latter attach to microcrystalline nuclei of calcium sulphate as they form, and hence increase their solubility. Soluble iron and manganese need to be removed if air is present in the system. If microbes are present, chlorine must be added, which with polyamide membranes means a subsequent dechlorination step with activated carbon or bisulphite. With careful pretreatment up to three years membrane life is possible, which is important as membrane replacement accounts for 15% of operating costs.

The other main format is based on polyamide hollow fibres. These are bunches of hollow tubes of outer diameter about 45 μm. The feed water is applied to the outside surfaces and the product water collected from the inside at the end of the fibre bunch. Pretreatment is more stringent as cleaning by mechanical means is not easy, although air scouring is quite feasible. It is possible to achieve one to two years life if pretreatments are done carefully.

**Electrodialysis.** Electrodialysis is a nonpressurized membrane technique which uses an electrical field to drive anions and cations to the appropriate electrode (McRae, 1983).

A stack of membranes, alternatively positively and negatively charged, is placed between the electrodes. These anion and cation selective barriers result in the solution between one pair of membranes becoming depleted in ions, while the compartments on either side become enriched in ions.

Each cell, comprising two membranes and two spacers, is 3 mm thick. They are arranged in stacks of up to 500 cells to give a stack height of 1.5 m, which will treat 200 kL/d. Pretreatments are simplified by reversing the polarity and the water flow every 20 minutes to flush out scale. Chemical cleaning is required every few weeks, and electrode replacement every 2 to 6 months.

The plants are more forgiving than reverse osmosis systems; the membranes last twice as long and are not prone to silica fouling, but the feed water should still have a low turbidity (Nadebaum, 1985). The power requirements are about half those of reverse osmosis and are more readily adapted to solar power. The choice of process depends on the character of the raw water, its cost (which makes product recovery important), energy considerations and chemical costs. A higher quality of water is possible with reverse osmosis, but operation is not as simple (Codina, 1987). Production costs for both processes from Australian operating data (Smith and Swinton, 1988) are similar at $3/kL for small plants run without skilled technical support in remote regions. For large scale units at power stations where expert attention is provided, reverse osmosis costs are lower, at less than $1/kL.

**SIROTHERM ion exchange.** Conventional ion exchange is a well established procedure for the preparation of high quality waters for boiler feed. Strong electrolyte resins of the sulphonic acid and quaternary ammonium type are generally utilized. Each requires at least a two-fold amount of acid or alkali for regeneration (relative to the stoichiometric quantity) which poses an effluent disposal problem and results in high operating costs. The method is too expensive to consider for potable water production.

CSIRO has developed the SIROTHERM process which employs a mixture of weak electrolyte resins; the mixture is not separated for regeneration, which is achieved with a mere hot water wash at 80 to 90$^\circ$C (Bolto and Weiss, 1977). These weak electrolyte resins are not always present in the charged form. The adsorption sites arise from the transfer of a proton from the acidic to the basic species:

$$\overline{HA} + \overline{B} + Na^+ + Cl^- \underset{hot}{\overset{cold}{\rightleftharpoons}} \overline{NaA} + \overline{BHCl} \tag{1}$$

The dissociation of the resins does not vary greatly with temperature, but the dissociation constant of water increases 30-fold from 25 to 85$^\circ$C, and the additional hydrogen and hydroxyl ions produced can be viewed as the regenerants. The process is used for partial demineralization of brackish waters of TDS up to 2000 mg/L to obtain a product water of TDS 500 mg/L. It is not used in its own right for making very high quality water for boiler feed.

There is a rate problem with such a process, but this was overcome by using very small resin particles of size 1 to 5 μm, with both types embedded within a composite bead of normal size (300 to 1200 μm). Continuous rather than batch operation has enormous economic advantages (Bolto, 1984). CSIRO has devised a continuous counter current system which makes use of magnetic composite beads which contain γ-iron oxide and are also reduced in size (200 to 300 μm). The resin is magnetized and left in that form (in contrast with the SIROFLOC procedure). As well as settling rapidly because of magnetic flocculation, the particles react rapidly since the flocs can be broken up on stirring. The high voidage of the flocs means that they can be directly pumped without great attrition, which is a unique feature not found with conventional resins.

The contactors devised for magnetic resins are sieve-plate columns, with a stirrer in each compartment (Swinton et al., 1983). Two columns are used; one is kept cold, the other hot. The cold feed water and the hot water regenerant enter from the bottom, and the resin moves downflow through each column. It is fluidized by the upflow of water and the agitation provided by the stirrers. Truly continuous operation is achieved with no flow interruptions. The system is quite different to other so-called continuous contactors, which are mostly intermittent in their operation.

Continuous magnetic SIROTHERM has been used to desalt 1 ML/d of a brackish water at Perth. The feed water is anaerobic, so no oxygen removal is necessary as would be the case with a surface water. The only pretreatment is softening. The magnetic resin contactor has hence been proven on the large scale. Although the raw water is warm at 40°C, an 80% yield of 500 mg/L product water is obtained from the 1400 mg/L raw water by regenerating the spent resin at the higher than usual temperature of 98°C.

**Costs of desalting brackish water.** SIROTHERM has fundamental advantages over competing membrane technology because it utilizes the cheapest energy source in the form of low grade heat, it removes the minor component rather than extracting the water, and it shows economies of scale since it does not involve a modular system. These advantages only become pronounced for large-scale plants, where the capital cost for a 40 ML/d plant is projected to be half that of reverse osmosis or electrodialysis (Gutteridge et al., 1983). The capital costs of the latter two techniques are similar for small-scale plants, but electrodialysis is more expensive for larger plants. The operating costs are similar, but SIROTHERM has an advantage if waste heat is available.

### Hardness Removal

Treatment for hardness is not carried out very much on the municipal scale in Australia, there being about 15 installations in existence (Gutteridge et al., 1983). For permanent hardness, where there is little bicarbonate present, lime treatment plus soda ash is employed. For temporary hardness, where the impurities are essentially calcium and magnesium bicarbonates, lime addition precipitates calcium carbonate and magnesia. A coagulant aid assists precipitation in these examples. The pH needs to be lowered afterwards, and this is usually achieved with carbon dioxide.

Cation exchange is often done on a household basis, especially in Adelaide. Excess brine is the regenerant for the sulphonic acid resin employed. However, these days the removal of calcium by replacement with sodium is not advocated for potable supplies on health grounds.

Dealkalization with a weak acid resin can be applied to bicarbonate waters, and is employed as a batch process in three different areas. Charged sites are formed on the resin by the reaction of acid groups with the bicarbonate. Calcium is adsorbed and carbon dioxide is evolved. Hence there is a TDS removal by the elimination of calcium (and magnesium) bicarbonate, without the addition of other ions to the product water:

$$\overline{2RCO_2H} + Ca^{2+} + 2HCO_3^- \rightleftharpoons \overline{(RCO_2)_2Ca} + 2CO_2 + 2H_2O \qquad (2)$$

The resin is regenerated with mineral acid, but at only about 5% over the stoichiometric amount.

The pioneering work on magnetic resin technology was actually carried out on the dealkalization reaction (Swinton et al., 1983). A shell graft resin of size 200 to 300 mm was synthesized for the purpose by grafting polyacrylic acid onto a crosslinked polyvinyl alcohol containing γ-iron oxide (Bolto, 1983). Because of its open structure, this 'whisker' resin reacts five times faster than conventional weak acid resins of the same size. Pilot plant work on a 50 kL/d scale operated in parallel with an existing batch plant showed that the magnetic resin inventory was 20% of that of the conventional resin in a batch system. An additional benefit is that no preclarification is necessary with the continuous magnetic fluidized bed system.

Magnetic resins are useful for treating slurries and sludges, and the removal of heavy metal ions from industrial effluents and sewage sludge with a magnetic sulphonic acid resin is currently under study. Other applications exist for the recovery of chemicals in biotechnology and hydrometallurgy.

### Taste and Odour Control

Taste and odour usually arise from algae and decaying vegetable matter. Algal growth can be prevented by limiting the nutrients present in effluents going to sensitive receiving waters, but this is difficult if fertilizers are the source. Copper sulphate may be added to reservoirs to minimize algal production. Dissolved air flotation can be used to remove algae from drinking water. However, the taste and odour compounds such as geosmin and 2-methylisoborneol, and even algal toxins, may still be present as soluble material. These give the water an earthy, swampy taste. Treatment with activated carbon is then necessary, as has been found for North Richmond water near Sydney (Nicholson et al., 1987).

Chlorine can aggravate taste problems, so disinfection is better carried out with chlorine dioxide or ozone. Ozonation may extend the life of the activated carbon, which may counterbalance the additional expense. Ozonation before the activated carbon column degrades the offending organics into fragments which are more easily biodegradable by the bacteria present in the column (Glaze, 1987).

### Nitrate and Fluoride

There are needs for the removal of these contaminants from waters in at least three States, but such treatment is not yet practised in this country. Nitrate in drinking waters is of current concern in Europe. Its removal may be accomplished by biological denitrification in anoxic systems. A carbon source must be provided for the bacteria, and methanol, acetic acid or ethanol are advocated (Solt, 1987). An alternative is to use hydrogen gas as the proton donor, and bicarbonate as the carbon source. These processes are the subject of full-scale studies; the water must be reoxygenated after treatment. For small-scale systems ion exchange is preferred, with bicarbonate as the regenerant.

Excessive amounts of fluoride can be removed by precipitation with lime if magnesium ions are present. The most common method is ion exchange using either bone char or alumina as the adsorbent (US Environmental Protection Agency, 1978). Regeneration is with caustic soda.

### Point-of-use Treatment

Household treatment at the point of use is an in-miniature version of conventional technologies. A good quality water for drinking and food preparation is the target, so units are generally located under the kitchen sink. For a small capital investment (up to $400 in the US in 1986) one may install reverse osmosis membranes for desalting or adsorbents such as alumina for fluoride, arsenic or selenium, activated carbon for organics or ion exchange resins for heavy metal ions or nitrate (Gumerman et al., 1986). Maintenance and operating costs are high, at an annual value roughly equivalent to the outlay for the original unit. It is not possible to optimize the processes though, and bacterial growth on the surfaces of the membranes or adsorbents is a problem. Smells are created by the dead bacteria, and disinfection is hard to achieve. End-of-tap units are being devised for this purpose. Point-of-use treatment is acceptable for isolated homesteads, but it is not economic where treatment can be provided on a town basis.

### WATER REUSE

Treated wastewater is a significant resource, at least for nonpotable uses. The immediate problems are more of promotion and economics. Clearly, improved wastewater treatment processes will cut the costs of recycling reclaimed water. Current reuse of secondary sewage effluent in Australia is 4-7%, but it could be raised to 40% as at least 100 urban centres have such potential (Gutteridge et al., 1983). Disinfection by ponding for 5 to 16 days or chlorination to the standard of no more than 1000 faecal coliforms/100 mL is required. Many towns in Western Australia, South Australia and New South Wales practice reuse for:

- irrigation of sports areas and parks, orchards, some crops and certain pastures,

- industrial purposes such as cooling water, minerals extraction and dust suppression,

- municipal applications as in road making and fire fighting.

Reuse is not practical for potable supply at the moment because of concerns over micropollutants, viruses, parasite eggs, reliability and cost.

Eventually the standard of treatment which can be attained at reasonable cost will approach that of potable water, when the additional treatment necessary is enhanced by the application of cheaper and better methods. Hence the possible future reuse of treated wastewater to augment freshwater supplies for potable purposes should be kept under continuing review.

The reuse of industrial wastewater is further advanced, where for economic reasons closed-loop systems for water conservation and the recovery of useful chemicals is advantageous. Physico-chemical methods are utilized for recovering inorganic and organic chemicals from aqueous systems, with recycling of the water (Bolto and Pawlowski, 1987).

The reduction of a person's daily water requirement of five litres to but a thimble full per day by recycling is heralded in science fiction for the 25th Century (Herbert, 1966). The Fremen of Arrakis with their stillsuits recirculate condensate by breathing in through the mouth and out through the nose, where a tube connects to catch pockets. Wastes are processed in thigh pads with salt precipitators and high efficiency filter systems; osmotic action and breathing motion provide the pumping force. It is hoped that the water industry can reflect at least a fraction of this vision in the more immediate future.


## CONCLUSIONS

Australia, the driest inhabited continent, has significant water quality problems. Depending on the region, most water quality problems are microbiological, or are caused by turbidity, colour, salinity or other dissolved chemicals. The presence of these impurities, either from natural causes or as a result of human activities, means that treatment is usually essential.

Techniques for contaminant removal are generally of a physicochemical nature. The most widespread needs are for cost-efficient clarification, decolourization and disinfection. Methods such as coagulation/sedimentation/filtration can be used in different configurations; there are also variants which either rely on magnetic particle technology or avoid chemical usage altogether. Choice of the disinfection method is dictated by the quality of the supply and the desirability of maintaining residual protection.

Eliminating dissolved contaminants is best achieved with membranes or by adsorption. For desalinating brackish water the optimum method depends on the characteristics of the raw water and the size of the operation, with a membrane method being the logical choice, but with adsorption processes based on magnetic particle technology having a predicted advantage for very large scale systems of the future. Taste and odour control can be a complex issue, but is generally related to the presence of plant nutrients; the use of activated carbon is necessary, with or without preoxidation of the organic compounds responsible.

Reuse of secondary sewage effluent for irrigation, industrial and municipal purposes is already practised in most States of Australia. Much greater exploitation of this low-grade resource is warranted. Recycling of both water and chemicals present in industrial effluents will become viable as new water treatment technologies are extended to effluent treatment.

# REFERENCES

Anderson, N.J., Blesing, N.V., Bolto, B.A., Dixon, D.R., Priestley, A.J. and Raper, W.G., 1981. Further scientific developments in the SIROFLOC process, in: "Proc. 9th Fed. Conv. Aust. Water & Wastewater Assoc.", Perth, pp 14-1 to 14-9.

Australian Water Resources Council., 1987. "Report on the use of slow sand filtration in Australia", Water Management Series No. 7, Aust. Govt. Pub. Service, Canberra.

Baker, M.N., 1981. "The Quest for Pure Water", Vol.1 2nd edition, Amer. Waterworks Assoc., Denver, p 358.

Bolto, B.A., 1983. Some new water purification processes based on polymers, Prog. Polymer Sci., 9:89-114.

Bolto, B.A., 1984. The development of desalination in Australia, Desalination, 50:103-114.

Bolto, B.A. and Pawlowski, L., 1987. "Wastewater Treatment by Ion Exchange", Spon, London.

Bolto, B.A. and Weiss, D.E., 1977. The thermal regeneration of ion exchange resins, in: "Ion Exchange and Solvent Extraction", J.A. Marinsky and Y. Marcus, eds., Vol. 7, Dekker, New York, pp 221-289.

Codina, O.J., 1987. Side-by-side comparison of RO and EDR, Ultrapure Water, 4(7):24-27.

Collins, A.G. and Johnson, R.L., 1985. Reduction of turbidity by a coal-aluminium filter, J. Amer. Waterworks Assoc., 77:88-92.

Crossley, I.A., 1983. Desalination by reverse osmosis, in: "Desalination Technology, Developments and Practice", A. Porteus, ed., Applied Science, London, pp 205-248.

Crowdus, F., 1984. System economic advantages of a low pressure, spiral RO system using thin composite membranes, Ultrapure Water, 1(1):29-30,33.

Curtin, C and McNeill, A., 1987. Upgrading drinking water quality in Victoria - what is needed and why, in: "Proc. 12th Fed. Conv. Aust. Water & Wastewater Assoc.", Adelaide, pp 330-337.

Drew, W.M., Phillips, S.F. and Wallis, I.G., 1987. A slow sand filtration demonstration plant at Pyalong, Victoria, in: "Proc. 12th Fed. Conv. Aust. Water & Wastewater Assoc.", Adelaide, pp 85-92.

Engineering and Water Supply Department of SA, 1980. "Anstey Hill Water Filtration Plant", Govt. Printer, Adelaide.

Garman, D.E.J., 1987. Water supplies for small communities, in: "Proc. 12th Fed. Conv. Aust. Water & Wastewater Assoc.", Adelaide, pp 78-84.

Garman, D.E.J., Woods, L.E. and Wade, A., 1983. "Water 2000: Consultants Report No. 7 - Water Quality Issues". Aust. Govt. Pub. Service, Canberra.

Glaze, W.H., 1987. Drinking water treatment with ozone, Environ. Sci. Technol., 21:224-230.

Gumerman, R.C., Burris, B.E. and Hansen, S.P., 1986. "Small Water System Treatment Costs", Noyes Data Corp., Park Ridge, pp 416-439.

Gutteridge Haskins and Davey, and Smith, R.C.G., 1983. "Water 2000: Consultants Report No. 10 - Water Technology Reuse and Efficiency", Aust. Govt. Pub. Service, Canberra.

Herbert, F., 1966. "Dune", Gollancz, London.

Huber, C.V., 1985. Direct filtration, in: "Innovation in the Water and Wastewater Fields", E.A. Glysson, D.E. Swan and E.J. Way, eds., Butterworth, Boston, pp 25-34.

James M. Montgomery Inc., 1985. "Water Treatment Principles and Design", Wiley, New York.

Kolarik, L.O. and Chin, C.T., 1977. Clear water from electrochemical filter, Water News No. 2, p 12.

Kolarik, L.O., Priestley, A.J. and Weiss, D.E., 1977. The SIROFLOC process for turbidity and colour removal, in: "Proc. 7th Fed. Conv. Aust. Water & Wastewater Assoc.", Canberra, pp 143-161.

Larsson, H.F., 1987. Continuous contact filtration, in: "Proc. 11th Fed. Conv. Aust. Water & Wastewater Assoc.", Adelaide, pp 102-109.

Loos, E.T., 1987. Experiences with manganese in Queensland water supplies, Water, 14(1):28-30,37.

McRae, W.A., 1983. Electrodialysis, in: "Desalination Technology, Developments and Practice", A. Porteus, ed., Applied Science, London, pp 249-264.

Nadebaum, P.R. and Fish, E.J., 1985. Considerations in the application of the SIROFLOC process for colour and turbidity removal, *in:* "Proc. 11th Fed. Conv. Aust. Water & Wastewater Assoc.", Melbourne, pp 230-238.

Nadebaum, P.R., 1985. Desalination of brackish water in Australia by electrodialysis, *in:* "Proc. 11th Fed. Conv. Aust. Water & Wastewater Assoc.", Melbourne, pp 273-281.

National Health and Medical Research Council, 1987. "Australian Water Resources Council. Guidelines for Drinking Water Quality in Australia", Aust. Govt. Pub. Service, Canberra.

Nicholson, C.G. Browning, R.C., Walshe, M.W. and Law, J.B., 1987. Upgrading of North Richmond water treatment works for algal taste and odour removal, *in:* "Proc. 12th Fed. Conv. Aust. Water & Wastewater Assoc.", Adelaide, pp 185-192.

Peters, R., 1986. Australian rivers are different. *Eng. Aust.*, 58(20):35.

Pigram, J.J., 1985. "Issues in the Management of Australia's Water Resources", Longman, Cheshire.

Smith, B.R. and Swinton, E.A., Desalination costs in Australia: a survey of plant operators, *in:* "AWRC Water Management Series", Aust. Govt. Pub. Service, Canberra, to be published.

Solt, G., 1987. Nitrate removal: a compromise solution, *in:* "Water Quality International" No. 1, pp 29-30.

Strom, A.G., Sweeney, P.C. and Craig, K.C., 1987. Direct filtration plants in Australia, *ibid.,* pp 181-188.

Swinton, E.A., Bolto, B.A., Eldridge, R.J., Nadebaum, P.R. and Coldrey, P.C., 1984. The present status of continuous ion exchange using magnetic micro-resins, *in:* "Ion Exchange Technology", Soc. Chem. Ind., London, pp 542-562.

Swinton, E.A., Nadebaum, P.R., Monkhouse, P and Poulos, A., 1983. Continuous ion exchange using magnetic micro-beads - field trials of a transportable pilot plant, *in:* "Proc. 10th Fed. Conv. Aust. Water & Wastewater Assoc.", Sydney, pp 30-1 to 30-14.

US Environmental Protection Agency, 1978. "Manual of Treatment Techniques for Meeting the Interim Primary Drinking Water Regulations", EPA-600/8-77-005, Cincinnati, pp 20-21.

Williams, P.G., Van Vuuren, L.R.J. and Van der Merwe, P.J., 1985. Dissolved air flotation upgrades a conventional plant, *in:* "Proc. 11th Fed. Conv. Aust. Water & Wastewater Assoc.", Melbourne, pp 189-196.

# THE ROLE OF SURFACE AND COLLOID CHEMISTRY

# IN THE SIROFLOC PROCESS

David R. Dixon and Luis O. Kolarik

CSIRO Division of Chemicals and Polymers
Clayton, Victoria

## INTRODUCTION

SIROFLOC is a water or wastewater treatment process developed at the CSIRO Division of Chemicals and Polymers' Laboratories in Melbourne. In common with all other treatment methods which seek to remove colour and turbidity from natural waters, the SIROFLOC process is an interfacial one dependent for its success upon what happens at the various interfaces present within the system. This is a review of the many laboratory and pilot plant studies which have been undertaken during the development of the SIROFLOC process with the intention of better defining the role of surface and colloid chemistry.

Magnetite added to a raw water can interact with many of the components typically present, not only the colour bodies and turbidity particles but also other soluble species both cationic and anionic and other colloidal matter. Examination of each possible two-component system indicates that the dominant interaction is between magnetite and the natural organic compounds. With this basis, a working model has been developed of both the clarification and regeneration stages of the process.

The SIROFLOC process for water clarification possesses a number of advantages over conventional water treatment methods for removing colloidal and other impurities from feedwaters. It offers greatly improved kinetics of clarification and sedimentation which translates into capital cost savings and most importantly avoids problems of sludge disposal. The highly concentrated effluent produced by regenerating the magnetite, is of low volume and its disposal is much easier than that of the highly gelatinous sludge from conventional plants.

## PROCESS OUTLINE

A clear picture of the basic steps in the process as currently practised at Mirrabooka and Bell Bay in Australia and Morehall in the U.K. can be obtained from the block diagram shown in Figure 1 (Dixon and Priestley, 1985).

Initially, the feedwater pH is adjusted to the optimum value for the process, which can vary from 5.0 to as high as 8.5 depending on the particular water, although in general, the lower the pH the better the performance. Subsequently, the water is dosed with freshly regenerated magnetite slurry at a rate which can vary from 0.5 to 2.0% w/w of the raw water flow. The magnetite must be demagnetized before addition, otherwise it clumps together and does not present its full surface area to the water. The

*Surface and Colloid Chemistry in Natural Waters and Water Treatment*
Edited by R. Beckett, Plenum Press, New York, 1990

demagnetizing happens automatically merely by pumping from the regeneration to the clarification stage.  The first period of contact time between water and magnetite can vary from 5 to 10 minutes, during which most of the colour and some of the turbidity is removed.  Prior to the second period, a cationic polyelectrolyte is dosed at a rate which can vary from 0.1 to 2.0 mg/L, depending mainly on the level of turbidity in the raw water.  Polyelectrolyte is used exclusively as the secondary flocculant.  During the second period, which can be from 2 to 4 minutes, the final traces of colour are removed and the polyelectrolyte firmly binds the remaining turbidity onto the magnetite surface.  Total contact time in the process is from 7 to 14 minutes.

The contact steps are followed by separating the magnetite from the clarified water, which has a turbidity generally less than 1 NTU and colour less than 5 Pt-Co units.  The magnetite is then cleaned and regenerated prior to reuse, and it is these steps which make the process unique.  The magnetite used in the SIROFLOC process, after regeneration with caustic soda, acts as a coagulant in its own right and eliminates the requirement for conventional coagulants such as alum or ferric chloride.  In regeneration there are two different steps.  Firstly, the turbidity and colour bodies are stripped off the magnetite and, secondly, the magnetite slurry is washed.  Regeneration requires a pH level between 11 and 12, while the washing step is usually carried out at a pH level between 10 and 11.

The SIROFLOC process relies on the ability of surface activated magnetite particles to destabilize and coagulate colloids and to adsorb colour bodies from the water.  Reuse of the magnetic adsorbent depends upon the efficiency with which the adsorbed or coagulated matter is released from the magnetite surface in regeneration.

As outlined in Table 1 the major components of the complex natural system are suspended particles of usually clay or silicate minerals, or biocolloids such as algae,

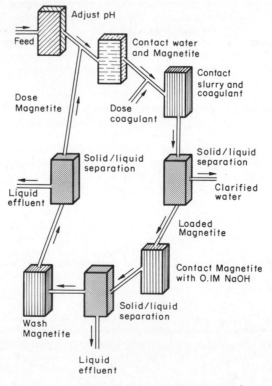

Figure 1.    SIROFLOC process block diagram.

Table 1.  Species Commonly Found in or Added to Natural Waters and their Interaction with Magnetite

| Component | Mechanism of interaction with magnetite | Significance |
|---|---|---|
| **SOLUBLE SPECIES** | | |
| Simple electrolytes | Electrostatic | Regeneration |
| Hardness cations $Ca^{2+}$, $Mg^{2+}$ | Adsorption | Treatment of hard feedwaters |
| Heavy metal ions $Pb^{2+}$, $Zn^{2+}$, $Mn^{2+}$, $Cu^{2+}$ | Adsorption | Metal ion recovery from effluents |
| Anions $PO_4^{3-}$, $SO_4^{2-}$, $SiO_4^{4-}$ | Adsorption | Removal of $PO_4^{3-}$ from from agricultural effluents |
| Organic acids humic and fulvic acids | Adsorption | Removal of colour from feed waters, fulvic acids and on occasions pesticides, surfactants and other man-made pollutants. |
| **SUSPENDED SOLIDS** | | |
| Clays, silica | Heterocoagulation | Removal of turbidity |
| Biocolloids algae, viruses, bacteria | Heterocoagulation | Algal harvesting, removal of pathogens |
| **ADDITIVES** | | |
| $Al^{3+}$, $Fe^{3+}$ salts | Adsorption | As secondary coagulants or surface coatings |
| Polyelectrolytes | Adsorption | Improved clarification and regeneration |

bacteria and viruses, and soluble species both inorganic and organic in character. There are the simple electrolytes ($Na^+$, $K^+$, $Cl^-$, etc), hardness cations ($Ca^{2+}$, $Mg^{2+}$), heavy metal ions and anionic species such as $PO_4^{3-}$. $SO_4^{2-}$, and $SiO_4^{4-}$, and the largely uncharacterized colour bodies comprising organic acids, known generally as the humic substances some of which may be of colloidal dimensions. The mixture may be further complicated by the presence of man-made pollutants in the form of pesticides, herbicides, surfactants etc. Furthermore, conventional water and wastewater treatment usually involves adding more surface active species in the form of aluminium or ferric coagulants and commercial flocculants. In the SIROFLOC process, magnetite of near colloidal dimension is used instead of the inorganic additives. Each component is either colloidal or surface active and in one way or another influences the interfaces present in the system. Thus the SIROFLOC process is very much within the realm of surface and colloid chemistry since both the adsorbent and the suspended particles are of colloidal dimensions and the soluble impurities are removed by adsorption at the magnetite surface. Optimization of the process is only achieved by understanding the mechanisms involved and this requires a detailed knowledge of what is happening at the magnetite-water interface. The approach adopted involved studying the effect of each component on the properties of the magnetite surface before proceeding to multi-component systems.

## EFFECT OF SIMPLE ELECTROLYTES

The effect of increasing ionic strength on the SIROFLOC process should be to assist clarification and hinder regeneration. The higher the concentration of a simple electrolyte, the lower the effective zeta potential on the turbidity colloids and the greater the driving force for coagulation of these particles. In regeneration, with increasing ionic strength, the release efficiency of the coagulated or adsorbed impurities decreases.

The level of total dissolved solids (TDS) in natural waters is largely determined by the rainfall, soil type, vegetation and land use of the catchment area. The best known examples of clarification of a water by increasing the TDS level are within river estuaries (Sholkovitz, 1976). With the SIROFLOC process there is no scope for controlling the ionic strength of the feedwater, however, there are decisions to be made regarding regeneration conditions such as pH and the choice of washwater which may be influenced by ionic strength considerations. Experiments were carried out in which the ionic strength or conductivity and the pH of the regenerant solutions were varied to observe the effects on colour and turbidity release with and without polyelectrolyte.

From results obtained in the absence of polyelectrolytes, it was found that the release of both colour and turbidity increased with increasing pH at the same conductivity and decreased with increasing conductivity at constant pH. The fact that turbidity release is more sensitive to increasing conductivity than colour release is suggestive that only some of the colour bodies can be considered as colloids. The presence of polyelectrolytes has little effect on colour desorption but does hinder turbidity release.

In mechanistic terms it seems that the smaller, soluble colour bodies adsorb on magnetite by forming complexes, presumably with the iron ions, and can be exchanged with hydroxide ions in regeneration. The higher the regeneration pH, the greater is the release. Some of the organics are colloidal in nature and their release from the magnetite surface is more difficult, requiring electrostatic repulsion which is hindered by higher ionic strength. The turbidity particles on the other hand, interact by heterocoagulation and their release depends upon both electrostatic repulsion and mechanical separation from the magnetite. It is the latter which is most affected by the use of polyelectrolytes.

Hence in a plant using a flocculant in clarification, regeneration should be performed at a sufficiently high pH to desorb colour despite the slightly harmful effects of increasing ionic strength. For turbidity release, the advantages of higher pH may be outweighed by the unavoidably higher ionic strength and therefore the wash water with the lowest possible conductivity should be used.

In practice the choice of regeneration pH at a plant is determined by both operational experience and economic factors such as the cost of alkali. The choice of wash water is also limited by the availability of suitable streams, but at the Mirrabooka plant when wash waters of different conductivities were compared, no differences in regeneration efficiency were observed. This apparent contradiction with laboratory results is attributed to the more vigorous agitation in the plant. Turbidity release is simply more dependent upon mechanical rather than electrostatic forces and is enhanced by the magnetic drum separators of the plant . This is a timely reminder of the difficulties encountered in extrapolating from a laboratory rig to a full-size plant.

## EFFECT OF HARDNESS

One of the more common components of natural waters is hardness, i.e. the presence of significant concentrations of calcium and magnesium ions. Although the

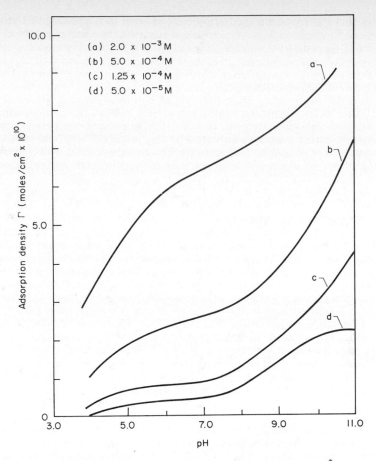

Figure 2.  Adsorption density-pH curves for the magnetite-Ca$^{2+}$ system for several different initial calcium ion concentrations.

effect of hardness on conventional clarification methods is minimal, it was considered that this might not be the case for the SIROFLOC process.

Basic adsorption theory (James, 1973) suggests that uptake of alkaline earth cations by an inorganic oxide such as magnetite should be preceded by at least partial hydrolysis. Even in preliminary experiments it was obvious that adsorption occurred in pH regions where the predominant metal ion species were unhydrolysed and the magnetite was positively charged, i.e. below the isoelectric point (IEP).  Since in this region adsorption cannot be due to electrostatic forces, it must result from a specific chemical interaction between the cations and groups present on the magnetite surface, leading to the formation of surface complexes.

Evidence for this was found in a combination of adsorption and electrokinetic experiments and the results shown in Figures 2 and 3 were obtained (Dixon, 1984, 1985). The data indicate uptake of the nonhydrolysed cations even at pH values equal to and lower than the IEP, indicating the occurrence of specific adsorption.  Consequently the surface charge of magnetite becomes more positive and even with relatively low cation concentrations charge reversal occurs at pH values above the original IEP of magnetite (pH 6.5).

From these results it is obvious that the presence of hardness in a feedwater to a SIROFLOC water clarification plant has significant effects.  At the normal clarification pH (about 6.0) for a feedwater with a normal degree of hardness there would be little change due to the divalent cations. However, a hard feedwater commonly has a natural pH value greater than 6.0, whereupon enhanced adsorption of Ca2+ and Mg2+ occurs, increasing the positive surface charge on the magnetite and aiding the removal of colour and turbidity. Furthermore, in regeneration, the high pH ensures complete adsorption of Ca2+ and Mg2+ ions which will certainly reduce the expected negative charge of the magnetite at this pH, and may, depending upon the level of hardness present, even cause charge reversal. Under these conditions, regeneration and release of the attached colloidal impurities would be extremely difficult. For successful plant performance, regeneration should be carried out in a soft environment, i.e. in the absence of Ca2+ and/or Mg2+. These predictions were confirmed and their relevance to the practical situation has been verified in both laboratory scale experiments and on a plant level (Anderson and Dixon, 1982).

## EFFECT OF HEAVY METAL IONS

Transition metal ions are sometimes present in natural waters, and hydrolysis is a major consideration with these ions. In the absence of specific adsorption the lower charged, less strongly hydrated, hydrolysis complexes have more favourable free energies

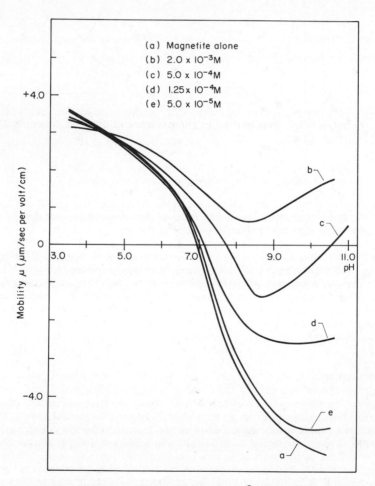

Figure 3.    Mobility-pH curves for the magnetite-Ca$^{2+}$ system for several different initial calcium ion concentrations.

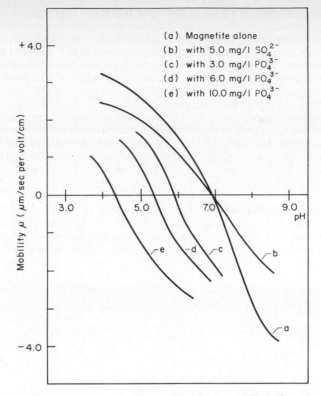

Figure 4.   Mobility-pH curves for magnetite (a) alone (b) in the presence of 5.0 mg/L $SO_4^{2-}$ ions and (c) in the presence of different concentrations of $PO_4^{3-}$ ions.

for adsorption at the oxide-water interface due to hydration energy effects (James and Healy, 1972).   Consequently, it is only in pH regions where the concentrations of these species are large that significant adsorption occurs.

The effect that the presence of low concentrations of heavy metal ions in the feedwater has on the removal of colour and turbidity was investigated.   For these experiments the water source was the Yarra River with colour of about 60 Pt-Co units and turbidity of about 14 NTU.   As water from this river is generally free of heavy metal ions at the sampling point, a spike of 1 mg/L of $Zn^{2+}$ was added.   Magnetite successfully removed colour and turbidity at low pH (about 5.0) in the presence of 1 mg/L $Zn^{2+}$, over several cycles of clarification and regeneration.

## EFFECT OF ANIONS

Many of the above considerations also apply to the adsorption of anions on inorganic substrates. Anions such as nitrate and chloride act as indifferent electrolytes at oxide surfaces, whereas phosphate anions are known to be chemically adsorbed on oxides (Stumm, 1981).   Referred to as specifically adsorbed anions, they are capable of lowering the IEP as a function of anion concentration (as shown in Figure 4).   The presence of sulphate ions reduces the magnitude of the positive zeta potential of magnetite but has little effect on the IEP, indicating that sulphate ions are not specifically adsorbed whereas phosphate ions are adsorbed strongly.

## ROLE OF ORGANIC ACIDS IN THE SIROFLOC PROCESS

The unique character of naturally occurring organic compounds has intrigued and frustrated researchers over several decades (Stevenson, 1982). These humic substances are the ubiquitous end-products of the partial degradation of plant and animal material. They are found not only in natural waters but also in soils, sewage and sediments.

Characterization and even standardization of stock supplies has proved to be very difficult. The natural organics are capable of chelating metal ions (Schnitzer, 1969; Gjessing, 1976), degrading when treated with disinfectants to form species such as trihalomethanes (Rook, 1974), absorbing pesticides and herbicides (Hayes, 1970; Adams, 1973), competing with other anions for adsorption sites (Davis, 1982) and acting as surfactants (Visser, 1964; Beckett, this volume). Thus in one way or another humic substances interact with all components of natural waters.

It is the latter property of surface activity which is of immediate interest to our study. The importance of organic coatings on particulate matter can best be illustrated by the electrophoretic data shown in Figure 5. In this experiment, the mobilities of three solids of vastly different character, bentonite, an impure alumina, and magnetite

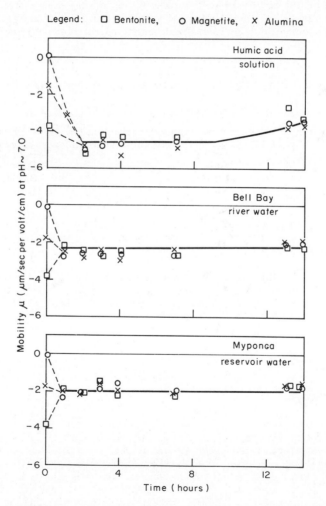

Figure 5.    Mobility-time curves of bentonite, alumina and magnetite dispersed in solutions of commercial humic acid, Myponga Reservoir water and Bell Bay river water.

dispersed in solutions of a commercial humic acid and in Myponga (South Australia) and Bell Bay (Tasmania) waters, were measured. The results demonstrate the ability of the organics in each water to dominate the surface properties, regardless of the nature of the original substrate. Similar studies on other waters have reached the same conclusion (Hunter and Liss, 1979; Beckett, this volume and references therein). Subsequent experiments of the same ilk but of far shorter duration produced similar results. The kinetic aspects of this interaction are the subject of a more comprehensive study (Yu et al., 1988).

It must be realized that all the processes occurring at interfaces within natural systems and during water treatment are affected by the presence of organic coatings. Obviously any laboratory attempt to model clarification of real waters must include humic material. This is true both for conventional coagulation with aluminium or ferric salts and for the SIROFLOC process.

In another series of experiments the surface properties of magnetite samples, pretreated in different ways, were investigated by streaming potential measurements in different solutions or raw waters. The magnetites used were a raw or untreated sample with residual clay impurities, a thoroughly clean sample, and two other samples with inorganic coatings of $MnO_2$ and $SiO_2$. The solutions used were prepared from commercially available fulvic acid and humic acid and natural waters from Myponga and Bell Bay. The results at two different pH values are presented in Table 2. The zeta potentials shown were calculated from the streaming potential measurements.

Table 2.    Effects of Humic Substances on the Surface Properties of Different Substrates

| AQUEOUS PHASE | MAGNETITE | ZETA POTENTIAL (mV) | |
|---|---|---|---|
| | | pH 6 | pH 8 |
| $10^{-3}M$ KCL | Clean | +25.5 | -14.6 |
| | Raw | -29.8 | -39.4 |
| | $MnO_2$ | - 8.3 | -22.1 |
| | $SiO_2$ | -15.3 | -37.1 |
| HUMIC ACID | Clean | -33.6 | -33.7 |
| | Raw | -33.7 | -36.2 |
| | $MnO_2$ | -25.4 | -26.4 |
| | $SiO_2$ | -36.6 | -36.3 |
| FULVIC ACID | Clean | -33.3 | -32.9 |
| | Raw | -34.6 | -38.7 |
| | $MnO_2$ | -25.4 | -27.4 |
| | $SiO_2$ | -32.0 | -34.7 |
| MYPONGA | Clean | -20.2 | -21.6 |
| | Raw | -22.1 | -22.2 |
| | $MnO_2$ | -18.6 | -17.7 |
| | $SiO_2$ | -22.5 | -21.4 |
| BELL BAY | Clean | -23.2 | -25.0 |
| | Raw | -24.9 | -25.4 |
| | $MnO_2$ | -18.4 | -19.3 |
| | $SiO_2$ | -25.1 | -27.5 |
| FULVIC ACID 50 mg/L $Ca^{2+}$ | Clean | -18.2 | -18.3 |
| | Raw | -20.6 | -20.5 |
| | $MnO_2$ | -13.8 | -14.2 |
| | $SiO_2$ | -19.1 | -19.7 |

The data illustrate the overriding effect of the organic coating on the colloids which ultimately cause them to have similar surface potentials regardless of the nature of the original substrate. The exception is $MnO_2$-coated magnetite which has a higher surface area (9.1 $m^2/g$ compared to 4.0, 2.7 and 4.8 $m^2/g$ for clean, raw and $SiO_2$-coated magnetites respectively). The other comparisons of interest are the less negative potentials of colloids in the preconcentrated raw waters than in the prepared solutions. Analysis of the feedwaters shows the presence of $Ca^{2+}$ (47.8 and 6.1 mg/L) and $Mg^{2+}$ (47.9 and 12.2 mg/L) ions in Myponea and Bell Bay waters respectively. This level of hardness is obviously sufficient to reduce the magnitude of the negative potential of the surfaces by complexation with groups in the adsorbed organic layer. The addition of 50 mg/L of $Ca^{2+}$ ions to the fulvic acid solution significantly decreased the magnitude of the negative zeta potential, as shown at the bottom of Table 2.

In a SIROFLOC plant treating a water of average composition, the surface properties of magnetite, in common with all natural particulates, will be controlled by the adsorbed film of organic material. Any considerations of the clarification mechanism must include the concept of a coated magnetite particle.

## INTERACTION OF MAGNETITE WITH TURBIDITY PARTICLES

Heterocoagulation of natural particulates and magnetite is an important part of the SIROFLOC process, which takes place during the premix stage before flocculant is added. Published studies of the interactions within mixed colloidal dispersions have involved particles such as clays and silica (Depasse and Watillon, 1970) and more recently latices of known and characterized dimensions (Bleier and Matijevic, 1977) but not magnetite. Future studies in our Laboratory will investigate the interactions between natural particulates and magnetite as a function of parameters such as pH, ionic strength and solids ratio, in the presence and absence of humic compounds.

Preliminary studies using both electrokinetic and optical measurements in a two component system (magnetite-bentonite) have highlighted the difficulties involved in obtaining reproducible data, although it is already apparent that ionic strength and pH are important parameters. It is obvious that further work should make use of better defined colloids. It may be necessary to eventually replace not only bentonite with latices of different particle size and surface characteristics, but also to substitute magnetic latices for magnetite. If successful, this could then be extended to multi-component systems, including humic compounds.

## REMOVAL OF BIOCOLLOIDS

MacRae and Evans (1983,1984) at the University of Queensland have demonstrated the ability of magnetite to adsorb and desorb a wide range of bacteria from waters or wastewaters. The extent of the interaction between magnetite and the bacteria depends upon factors such as pH, ionic strength, hardness, competing colloids, agitation, polyelectrolyte dose, solids ratio and magnetite pretreatment. With three adsorption stages, the numbers of bacteria in a suspension could be reduced by five orders of magnitude, provided there is sufficient adsorption capacity to remove all particulate matter.

MacRae (1985,1986) also showed that by careful choice of the type of microbe adsorbed on magnetite, it was possible to remove unwanted components such as pesticides and herbicides from waters. They have removed lindane, 2,4-D and 2,4,5-T, and a series of other chlorinated hydrocarbons using the bacterium *Rhodopseudomonas sphaeroides* immobilized on magnetite. There are other projects under way invoking the same concept of utilizing magnetically immobilized microorganisms to achieve the removal of specific compounds.

At the same time as the above work with bacteria, Atherton and coworkers (1983), also at the University of Queensland, undertook research on the interaction of magnetite with some of the more common viruses found in natural waters. Because of their small size and the outer protein coating, viruses behave as charged colloids in aqueous suspension. Initial studies involving bacteriophage MS2 showed magnetite to be an efficient adsorbent, dependent upon the usual solution parameters.

Experiments with radiotracers showed that the virus which is adsorbed at for example pH 6, is disrupted at pH 10 when adsorbed to magnetite. In contrast, the free virus retains its infectivity at pH 10. This disintegration is of considerable significance as it means that magnetite as well as efficiently removing viruses from waters, on regeneration with alkali will release inactive material.

Similar results were obtained with the human pathogenic poliomyelitis virus and adenovirus and reovirus. An understanding of the adsorption of virus particles to magnetite and the effect of other components such as ions, clay and polyelectrolyte, requires a knowledge of the IEP of the virus concerned. The development of methods for determining the IEPs of bacteriophages and animal viruses has been reported (John, 1987).

## USE OF INORGANIC COAGULANT AIDS

For feedwaters of higher turbidity, it is necessary to enhance the capacity of magnetite as a coagulant-adsorbent by adding a secondary coagulant or flocculant. Usually a low dose of cationic polyelectrolyte is added, but one alternative is to use aluminium ions. Electrokinetic data confirm that hydrolysis products of these ions adsorb strongly on magnetite depending on both pH and concentration. The nature of the anion is an additional parameter which markedly affects the surface properties of the system.

As shown in Figure 6, sulphate, which has little effect on the surface properties of magnetite, is potential determining for the magnetite-$Al^{3+}$ system. A comparison of the effect of equivalent amounts of alum and for example aluminium chloride or nitrate on the surface properties of magnetite showed that much of the increase in (positive) zeta potential achieved by adsorption of aluminium hydrolysis products is negated by the subsequent adsorption of sulphate ions. It appears that sulphate ions specifically adsorb on $Al(OH)_3$ coated magnetite, reducing both the positive zeta potential and the IEP.

However, in jar tests with river waters the above effect of sulphate ions was not found, presumably due to the presence of the more strongly adsorbed organic anions. Again this is a reminder of the added complexity of the natural system. It was also apparent that aluminium ions only function as secondary coagulants at concentrations sufficient to form significant amounts of hydroxide flocs, not at the 1-2 mg/L level where they increase the positive zeta potential of the magnetite.

## USE OF POLYMERIC FLOCCULANTS

Cationic polyelectrolytes were introduced into the SIROFLOC process as an alternative secondary coagulant to alum to cope with higher turbidity feed waters. Having assisted in clarification, the polyelectrolyte must be such that it does not hinder in regeneration. Research to date has concentrated on choosing the best polyelectrolyte, commercial or otherwise, on the basis of continued clarification efficiency over a long period in both jar test experiments and pilot plant operations. Experience has shown that in general, chain length is an important consideration in turbidity removal and that charge density is more important for colour removal. The specific structure of the polyelectrolyte is also critical, particularly in regard to regeneration performance (Anderson et al. 1987).

The effect that cationic polyelectrolytes have on the surface properties of magnetite as a function of concentration, type of polyelectrolyte and pH was studied with a modified streaming potential technique. The colloidal magnetite particles were held in the form of a plug by external magnets and the streaming potential measured as a function of fluid flow, induced by applied pressure. The magnetically held plug was found to obey the requirements of the streaming potential technique. Before each measurement, there was a period of equilibration during which the solution was forced back and forth through the magnetite plug. The kinetics of adsorption of the polyelectrolyte were not as fast as anticipated and longer equilibration times were needed.

As shown in Figure 7, treatment of magnetite with a low dose of different cationic polyelectrolytes increased the IEP. For the strongly basic polymers of the epichlorhydrin/amine type with increasing molecular weight in the order C573 < C577 < C581, this was significant. With C460, another strongly basic polymer but of the acrylamide/ester type the increase is less marked. And with C521, a weakly basic polymer of low molecular weight and charge density, there is only a very slight change in the IEP.

Regeneration of the magnetite plug by treating it with alkali (0.1M NaOH for 10 minutes) lowered the IEP from the values measured after polyelectrolyte adsorption, as shown in Figure 8. For C573, C577, C581 and C521 (0.2 mg/L/g of magnetite) this decrease upon regeneration was significantly less than the increase observed after adsorption, suggesting that not all of the polymer was desorbed during regeneration.

However, with C460, the IEP of the alkali-regenerated magnetite was much lower than that of the original magnetite sample. This could only be due to the presence of an anionic impurity on the surface and suggests that the effect of alkali has been to hydrolyse the polyacrylamide leaving carboxylate groups which have lowered the IEP.

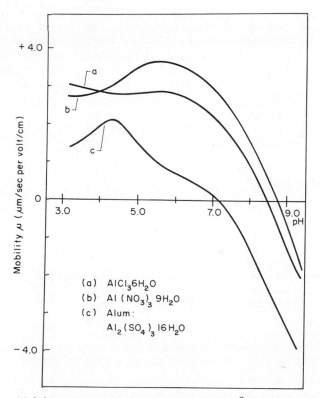

Figure 6.    Mobility-pH curves for the magnetite-$Al^{3+}$ system using 10 mg/L of different aluminium salts.

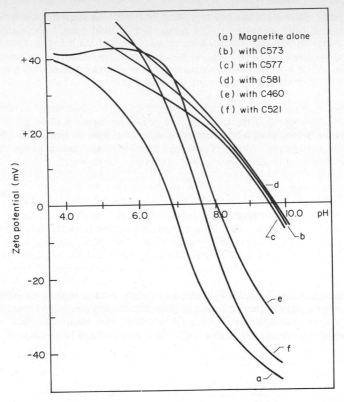

Figure 7.    Zeta potential-pH curves for magnetite equilibrated with different polyelectrolytes at a concentration of 0.2 mg/L/g of magnetite.

As might be expected, increasing the time of contact of loaded magnetite with 0.1M alkali, lowers the IEP, although there seems to be little further decrease after 10 minutes contact and as noted above some of the polymer remains on the surface. The extent of desorption also increases with increasing pH in an almost exponential manner.

The relatively rapid rate at which the original surface is restored or nearly restored is unexpected. Since desorption of a long chain polyelectrolyte is a cooperative event (Grant, 1975) it should be a much slower process than adsorption which involves a sequence of steps. That this is not observed highlights our lack of knowledge of both processes. It may be that some analogy with the magnetite-$Ca^{2+}$ system is possible, invoking the concept of specific adsorption due to the formation of some surface complex with Fe ions. Subsequent to this rapid step there is a slower process, more akin to the theoretical model of polymer adsorption. In regeneration, hydroxide ions may dissolve the outer layer of the magnetite surface, releasing the surface complex and partially restoring the original magnetite surface. Such a mechanism would not require a cooperative unzipping of the polymer from the surface and is consistent with the observed rate of desorption.

**OVERALL MODEL**

On the basis of the above data and from experience at both pilot plant and plant level it is very apparent that the humic compounds have the greatest influence on the interfaces present within the system and therefore on the efficiency of the SIROFLOC process.

The natural organics have the ability to interact with all the other components of the system, determining not only the character of the solids, both the natural turbidity particles and the added magnetite, but also the nature of the soluble species. Consequently in any future work with multicomponent systems, humic compounds will be included. While each of the other components has some influence on the efficiency of the process, their effects are minor when compared to that of the humic compounds.

With this background, we have attempted to develop a model of the SIROFLOC process as currently practised. In a typical raw water of average colour and turbidity and without significant levels of either alkaline earth or transition metal ions, there will be present a range of organics, both humic and fulvic types, and a range of particulate matter, coated with these organics.

In the first contact stage, the freshly regenerated magnetite will adsorb most of the soluble organics and some of the colloidal matter, thereby acquiring a negatively charged surface. The cationic flocculant, which is then added interacts with the rest of the soluble organics, the remaining organic coated particles and the partially loaded magnetite particles. The polyelectrolyte acts as a bridging agent, and provides the additional adsorption capacity needed for adequate clarification.

Efficient regeneration depends on reversing the above interactions, adding alkali provides a large concentration of a strongly competing anion, which exchanges with the more soluble organic species on the magnetite surface. Release of the colloidal matter is more difficult and requires assistance from both the mechanical and magnetic forces of the rotating magnetic drum separators and the hydrodynamic forces of the washing

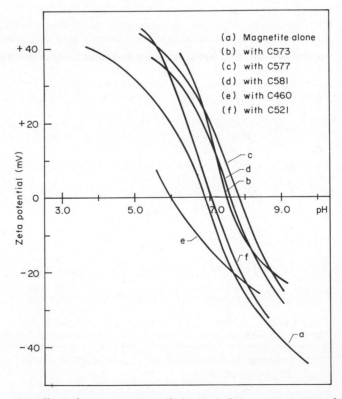

Figure 8.    The effect of regeneration with 0.1M NaOH on zeta potential-pH curves for magnetite previously equilibrated with different polyelectrolytes at a concentration of 0.2 mg/L/g of magnetite.

stages.  The presence of the polyelectrolyte in regeneration is a hindrance and this fact must be taken into account when selection of the dose and type of flocculant is being made.

The presence of other strongly adsorbing components such as calcium ions or transition metal ions or anions such as phosphate in the raw water will alter the above simplistic model.


## CONCLUSIONS

In summary each of the studies of the two component systems described above has reaffirmed the original premise that water clarification by the SIROFLOC process is an interfacial phenomenon, and that mechanistic understanding can best be increased by further colloid and surface chemical investigations.  Application of the results of such work to the operation of full-scale plants must be tempered by an appreciation of the difficulties of extrapolation and of the complexities of the real system.  It is hoped that future work will enable the development of a more detailed model.


## ACKNOWLEDGEMENTS

The authors readily acknowledge the assistance of the other members of the CSIRO Water Group in this work, together with the efforts of AUSTEP, the commercial licensee of the SIROFLOC process, and also the encouragement given by staff from the following water authorities and boards, the Water Authority of Western Australia, the Rivers and Water Supply Commission of Tasmania, and in the UK, the Water Research Centre and the Yorkshire Water Authority.

## REFERENCES

Adams, R.S., 1973. Factors affecting soil adsorption and bioactivity of pesticides, *Residue Rev.*, 47:1.

Anderson, N.J., Blesing, N.V., Bolto, B.A. and Jackson, M.B., 1987. The role of polyelectrolytes in a magnetic process for water clarification, *React. Polymer Ion Exch. Sorbents*, 7:47.

Anderson, N.J. and Dixon, D.R., 1982. Water clarification, *Aust. Patent 82365/82*.

Atheron, J.G. and Bell, S.S., 1983. Adsorption of viruses on magnetic particles, *Water Res.*, 17:943.

Atherton, J.G. and Bell, S.S., 1983. Adsorption of viruses on magnetic particles, *Water Res.*, 17:949.

Bleier, A. and Matijevic, E., 1977. Heterocoagulation, *J. Chem. Soc. Faraday Trans. I.*, 74:1346.

Davis, J.A., 1982. Adsorption of natural dissolved organic matter at the oxide-water interface, *Geochim. Cosmochim. Acta*, 46:2381.

Depasse, J. and Watillon, A., 1970. The stability of amorphous silicas, *J. Coll. Interface Sci.*, 33:430.

Dixon, D.R., 1984. Colour and turbidity removal with reusable magnetite particles, *Water Res.*, 18:529.

Dixon, D.R., 1985. Interaction of alkaline earth metal ions with magnetite, *Coll. Surf.*, 13:273.

Dixon, D.R. and Priestley, A. J., 1985. "The SIROFLOC Process for Water Clarification and Decolourization", CSIRO Division of Chemical and Wood Technolology Research Review, 13, Melbourne.

Gjessing, E.T., 1976. "Physical and Chemical Characteristics of Aquatic Humus", Ann Arbor Science, Ann Arbor, Michigan.

Grant, W.H., Smith, L.E. and Stromberg, R.R., 1975. Adsorption and desorption rates of polystyrene on flat surfaces, *J. Chem. Soc. Faraday Disc.*, 59:209.

Hayes, M.H.B., 1970. Adsorption of triazine herbicides on soil organic matter including a short review on organic matter chemistry, *Residue Rev.*, 32:131.

Hunter, K.A. and Liss, P.S., 1979. The surface charge of suspended particles in estuarine and coastal waters, *Nature*, 282:823.

James, R.O. and Healy, T.W., 1972. Adsorption of hydrolyzable metal ions at the oxide-water interface, *J. Coll. Interface Sci.*, 40:42.

John, M.A., 1988. Ph.D. Thesis, University of Queensland, St Lucia.

MacRae, I.C. and Evans, S.K., 1983. Factors influencing the adsorption of bacteria to magnetite in water and wastewater, *Water Res.*, 17:271.

MacRae, I.C. and Evans, S.K., 1984. Removal of bacteria from water by adsorption to magnetite, *Water Res.*, 18:1377.

MacRae, I.C., 1985. Removal of pesticides in water by microbial cells adsorbed to magnetite, *Water Res.*, 19:825.

MacRae, I.C., 1986. Removal of chlorinated hydrocarbons from water and wastewater by bacterial cells adsorbed to magnetite, *Water Res.*, 20:1149.

Rook, J.J., 1974. Formation of haloforms during chlorination of natural waters, *Water Treatment Exam.*, 23:234.

Schnitzer, M., 1969. Reaction between fulvic acid, a soil humic compound and inorganic soil constituents. *Soil Sci. Amer. Proc.*, 33:75.

Sholkovitz, E.R., 1976. Flocculation of dissolved organic and inorganic matter during the mixing of river water and seawater, *Geochim. Cosmochim. Acta*, 40:831.

Stevenson, F.J., 1982. "Humus Chemistry", Wiley, New York.

Stumm, W. and Morgan, J.J., 1981. "Aquatic Chemistry", Wiley, New York.

Visser, S.A., 1964. Oxidation-reduction potentials and capillary activities of humic acid, *Nature*, 204:581.

Yu, D., Fude, I., Dixon, D.R. and Priestley, A.J., 1988. A Kinetic Study of Turbidity Removal from Thai Waters by SIROFLOC Process, *Asian Environ.*, 10:71.

# COAGULATION AND FLOCCULATION - DESTABILIZING PRACTICES?

## (WITH PARTICULAR REFERENCE TO METAL ION COAGULANTS)

Peter R. Hutchison[1] and Thomas W. Healy[2]

[1]Iron Chemicals Division
Tioxide Australia Pty. Ltd.
Burnie, Tasmania

[2]Department of Physical Chemistry
University of Melbourne
Parkville, Victoria

## INTRODUCTION

In the production of clarified water for potable and industrial requirements, the role of surface and colloid chemistry provides the mechanisms and reactions upon which the unit operations for practical engineering are based. The complexity of this role cannot be underestimated because some mechanisms and reactions remain the subject of conjecture and continued investigation, thus attempts to advance comprehensive theories have not been made here.

Nevertheless there are predominant reactions and mechanisms that occur which determine our approach to water treatment practices. This paper attempts to provide an understanding of the reactions which contribute to the behaviour of our waters and in particular how destabilization mechanisms are utilized in practice.

Coagulation and flocculation are processes universally employed in the water treatment industry. Although loosely interchanged by many, a distinction has been made here between the two terms. The majority of components requiring removal in water purification processes are of colloidal size and present as a stable dispersion. Since Van der Waals forces tend to lead to agglomeration (i.e. flocculation) of particles, stability of the dispersion requires particles to repel each other which they do by carrying a nett electrostatic charge. Two groups of colloids (hydrophilic and hydrophobic) commonly encountered in water treatment, have been described below,as has the origin of the surface charge and its contribution to the stability of these colloidal dispersions. This has been the subject of continual review (Overbeek, 1977; Lyklema, 1981) and in addition much attention has been given to the characterization of colloidal components and the determination of the functional groups present, particularly in the hydrophilic organics which contribute to colour and haloform formation during chlorine disinfection processes (Sinsabaugh et al., 1986; Christman et al., 1983). In this paper the mechanisms for destabilization of hydrophilic colloids have been reviewed and the observation made that the total coagulant dose must account for the inherent plus precipitated turbidity loads. Investigations by Bursill et al. (1985) and Sinsabaugh et al. (1986) into the effectiveness of colour removal by coagulation show that the nature of the organics has a bearing on treated water quality. Examination of the mechanisms of destabilization of hydrophobic colloids present in natural waters reveals that the ultimate requirement to allow particle coagulation is for charge neutralisation, but the mechanism is dependent on such factors

*Surface and Colloid Chemistry in Natural Waters and Water Treatment*
Edited by R. Beckett, Plenum Press, New York, 1990

119

as colloid type, particle concentration, pH and coagulant concentration. The importance of the rate of adsorption is highlighted by the concept of "adsorption destabilization" as advanced by Bratby (1980).

The nett effect of these factors is to delineate zones of destabilization and restabilization (coagulation domains) and those that are expected in practice are detailed after Hutchison et al. (1985). Anions present are shown to influence coagulation by extending the range of optimum destabilization pH (sulphate) or by shifting the optimum destabilization pH (phosphate). The action of metal ion coagulants as a primary participant in coagulation/flocculation processes is linked to their ability to form multicharged polynuclear complexes with enhanced adsorption properties in aqueous systems. Choice is generally a balance of cost and effectiveness.

In this paper the stages of flocculation are reviewed. Efficient and intimate coagulant contact has been highlighted as the key parameter during the first perikinetic flocculation stage. Time and velocity gradient factors determine the rate and extent of particle aggregation during the next step designated the orthokinetic flocculation stage. The work of Mitsuo (1977) has been highlighted in promoting a third designated stage, mechanical syneresis, which is akin to a dewatering stage promoted by closer particle contact. Thus the many surface and near surface reactions in the colloidal realm play an important role in shaping the nature of our waters and water treatment processes.


## COAGULATION AND FLOCCULATION

An absolute definition of either terms will not be attempted, suffice to say that because of their wide ranging use and understanding, particularly in the water treatment industry free interchange of both terms has been the feature of many a discussion. In an effort to standardize thinking, in recent times, there has been a general acceptance of the following differentiation between coagulation and flocculation which will be the basis for terminology used in this paper:

**Coagulation** is the process by which components in a stable suspension or solution are destabilized by overcoming the forces which maintain their stability.

**Flocculation** is the process by which destabilized particles join together to form large stable particles or agglomerates.

This distinction embraces earlier suggestions by La Mer and Healy (1963) which describes the action of simple electrolytes as coagulation and the action of polyelectrolytes as flocculation; this earlier definition implies, correctly we believe, that ferric iron and aluminium salts for example act in part via a coagulation-flocculation mechanism.

Before going on to discuss these two processes it is important to understand the nature of these stable systems in the various types of waters which are encountered in water treatment practices.


## COLLOIDS AND SURFACES

Natural waters are composed of suspensions or dispersions and solutions of a wide variety of components. Emphasis is given in this paper to particles generally in the size range of $10^{-1}$ to $10^{-6}$ mm. (It is not unusual in water treatment to encounter particles up to 5 mm diameter after coarse screening, but these particles are generally removed by entrapment in whatever process is chosen). The presence and effects of soluble species will be dealt with where their influence on surface reactions is significant.

Suspended particles or dispersions of particles in the range of $10^{-1}$ to $10^{-3}$ mm are considered the macro particles of water treatment processes whilst those in the range of $10^{-3}$ to $10^{-6}$ mm are considered colloidal particles. The presence of two distinct phases then distinguishes colloidal suspensions from true solutions. It is at the interface between these two phases that the properties of the colloidal dispersions to be considered are evident.

By the very nature of their size colloidal particles exhibit large surface area to weight ratios and it is these large surface areas that contribute ultimately to the stability of their dispersions.

Historically a distinction has often been made between two types of colloidal "solutions" or sols, lyophobic and lyophilic which in aqueous systems are given the terms hydrophobic (literally water hating) and hydrophilic (literally water loving). The primary differences between the two types of sols, are listed in Table 1.

Most natural waters contain species of both types of colloids, therefore destabilization mechanisms and reactions are determined by the type and predominance of these species. The most popularly believed dominant mechanism for destabilization of hydrophobic colloids being a reduction of the nett surface charge to a point where colloid particles, previously stabilized by electrostatic repulsion, can approach closely enough for London/van der Waals forces to hold them together and allow aggregation.

The origin of the surface charges contributing to colloid stability is generally related to adsorption properties of the surface (Stumm and Morgan, 1962). Other factors which can contribute to surface charge are charge imbalance resulting from lattice imperfections at the solid surface as exhibited by some clays, and chemical reactions at the surface. These reactions can either be ionisation reactions with functional groups or coordinate bonding of solutes to the solid surface.

In the realms of water treatment four main processes are of interest:

- adsorption of organic solutes on organic and inorganic surfaces
- adsorption of inorganic solutes on organic and inorganic surfaces.


Table 1.    Principle Distinctions Between Hydrophobic and Hydrophilic Sols

| Hydrophobic Sols | Hydrophilic Sols |
|---|---|
| 1.  Surface tension similar to pure water | 1.  Surface tension often lower than water |
| 2.  Viscosity similar to water | 2.  Viscosity much higher than water |
| 3.  Small quantities of electro-lytes added will cause aggregation | 3.  Small quantities of electro-lytes have little effect. Large quantities may cause salting out. |
| 4.  Particles migrate in one direction in an electric fiel | 4.  Particles may migrate in either direction or not at all in an electric field |
| Examples | Examples |
| Metals, ores - metal oxides and sulphides, sulphur, silver halides | Gelatin, starch, gums, proteins, — viruses, bacteria |

In natural fresh waters (thereby implying low electrolyte concentration) we will assume the majority of colloidal particles encountered to be nonporous. The above adsorption reactions will result in the formation of a monolayer of ions of opposite charge to that of the colloidal solid, because counter-ion adsorption predominates over co-ion adsorption. This is known as the Stern layer. Outside this layer, a mobile layer will form by electrostatic attraction. This is known as the diffuse layer and consists of both positively and negatively charged species having a nett charge equal and opposite to the surface plus Stern layer charge.

These layers of charge are collectively called the electrical double layer and can be represented conceptually as in Figure 1. Most commonly it is these outer layers of diffuse charge which by electrostatic repulsion maintain stability of the colloidal dispersion.

If such particles are subjected to an applied electric field the particles will exhibit electrophoresis because displacement of the two layers with respect to each other will occur. Because of this the potential difference between the plane of shear and the bulk solution, known as the zeta potential, can be derived by a simple relationship. Most often a derived quantity such as the measured electrophoretic mobility is sufficient to characterize this electrostatic effect.

Zeta potential is a useful tool in providing information about the nature and stability of the surfaces and in the study of adsorption of species from solution. It can also provide a useful empirical means of water treatment process control providing correlation between practical testing, such as jar tests, and changes in raw water properties has been made.

There is some conjecture as to whether the principal components causing organic colour in natural water and some wastewater, mainly fulvic, hymatomelanic and humic acids, are colloidal or in true solution. However their large molecular size (either due to high MW or aggregation) and the presence of functional groups cause these compounds to exhibit some hydrophilic colloidal properties. (See Chapter by Beckett).

These colloids by definition have a strong affinity for water and may in some cases be bound with up to ten times their own weight of water. Also the presence of ionic functional groups resulting from ionisation of adsorbed species has been demonstrated (Black and Christman, 1963) although in many instances the effect does not extend beyond the hydrated layers. Hydrophilic colloids owe their stability to a combination of

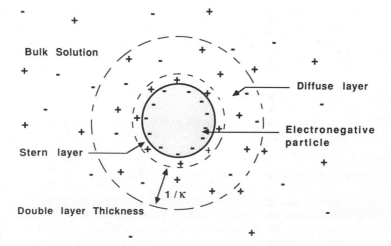

Figure 1.  Schematic representation of a negatively charged colloid particle with its associated electrical double layer. $1/\kappa$ represents the diffuse layer thickness or Debye-Huckel length (After Priesing, 1962).

this extensive hydration and the usually negative colloid charge contributed by functional groups present.

## DESTABILIZATION - COAGULATION OF HYDROPHILIC SOLS

Because of the hydrated layers present, destabilization by repression of the electrical double layer cannot be considered, even though hydrophilic colloids do carry a colloidal charge. This charge is in many cases confined within the hydrated layer. The negative colloidal charge of hydrophilic colloids generally results from the ionisation of ionogenic groups which in the case of organic colour consist principally of carboxyl and aromatic hydroxyl groups (Hine and Bursill, 1984).

For a variety of reasons there is a desire to remove organic colour in potable and industrial water production. These include aesthetic, taste, disinfection and industrial process considerations as well as their ability to form harmful haloforms by reaction with chlorine (Christman et al., 1983). Although, Sinsabaugh et al. (1986) showed that the removal of fulvic acids, the most dominant precursors in untreated waters, did not eliminate the presence of precursors for haloform formation in treated waters.

Destabilization of organic colour follows in a similar manner to hydrophilic colloids in general and is detailed by Stumm and Morgan (1962) who demonstrated the formation of complexes of metal coagulant ions with functional groups resulting in precipitation to form multiligand complexes with hydroxyl ($OH^-$) and water species. This metal ion reaction with hydrophilic colloids usually leads to precipitation of the colloid into a form that more closely resembles a hydrophobic colloid, which must then be destabilized along with other true hydrophobic colloids. The polymeric hydrophilic colloidal material that usually contributes to colour and taste can be thought of as producing a "coagulant demand" that must be satisfied before free coagulant is available for destabilization-coagulation. Thus the total coagulant dose must account for both the inherent plus the precipitated turbidity loads.

In most water systems containing polymeric or hydrophilic colloids it is fortunate that the metal precipitates and the inherent turbidity (hydrophobic colloids) respond in a similar fashion to the metal salts added in excess of the demand provided by the colour etc. species. Donnan et al. (1981) illustrated the nature of such demand by soluble polymeric and colour species in their work on paper mill effluent coagulation. This relationship is supported by the stoichiometric relationship between colour intensity and metal ion coagulant dosage for a particular water/colour (Narkis and Rebhun, 1977).

Ferric iron salts generally have been found to yield lower residual colour than aluminium salts primarily because of the stronger complexation with iron. Also the optimum colour removal pH for iron salts is lower than for aluminium. Lower pH values are required to limit the opportunity of the metal ion hydrolysis reaction and promote the functional group ligand reaction.

Black et al. (1963) and also others have shown a direct relationship between water colour intensity and pH depression for efficient destabilization. However, with this and other relationships with variables in water treatment processes it is important that correlation prediction be confirmed with practical procedures such as jar testing.

## DESTABILIZATION - COAGULATION OF HYDROPHOBIC SOLS

We have observed how the presence of small quantities of electrolytes can result in stability of hydrophobic sols and in some instances contribute to stability of hydrophilic sols and have discussed the importance of adsorption at the colloid surface in determining the surface charges which lead to this stability.

Quantitative expression of the theory of colloid stability known as the DLVO theory, predicts the effect of increased electrolyte concentrations on colloid stability due

Table 2. Characteristics of Destabilization Mechanisms with Metal Coagulants

| PARAMETER\MECHANISM | Influence of Indicated Parameter According to Mechanism | | | |
|---|---|---|---|---|
| | PHYSICAL DOUBLE LAYER | ADSORPTION LAYER | BRIDGING DESTABILIZATION | PRECIPITATION |
| Electrostatic Interactions | Predominant | Important | Subordinate | Subordinate |
| Chemical Interactions and Adsorption | Absent | Important | Predominant | May occur but not essential for removal |
| Zeta potential for optimum destabilization | Near zero | Not necessarily zero | Usually not zero | Not necessarily zero |
| Addition of excess coagulant | No detrimental effect | Restabilization usually accompanied by charge reversal; may be blurred by precipitation | Restabilization due to complete surface coverage | No detrimental effect |
| Fraction of surface coverage ($\theta$) for optimum floc formation | Negligible | $0<\theta<1$ | $0<\theta<1$ | Unimportant |
| Relationship between optimum coagulant dosage and particle concentration (for a given suspension) | Optimum dosage virtually independent of colloid concentration | Stoichiometry possible but does not always occur | Stoichiometry between dosage and particle concentration | Optimum dosage virtually independent of colloid concentration |
| Physical properties of flocs produced | Dense, great shear strength but poor filtrability in cake filtration | Flocs of widely varying shear strength and density | Flocs of 3-dimensional structure; low shear strength but excellent filtrability | Flocs of widely varying shear strength and density |

to a decrease in the thickness of the electrical double layer. We have discussed how zeta potential is a measure of the potential within the double layer and it therefore follows that colloid stability is closely related to the zeta potential of the particles. It is expected that by shielding the surface charge of the double layer by addition of electrolyte, viz metal ions, i.e. to reduce the zeta potential to zero, destabilization will occur.

However, in the coagulation of aqueous colloidal dispersions with metal ion coagulants it is observed on occasions that the relationship between zeta potential and colloid stability is not immediately obvious. This is because of the role that adsorption plays in destabilization reactions and was advanced by Bratby (1980) as "adsorption destabilization" which has the ultimate effect of reducing the effective surface charge hence reducing repulsive forces between particles and allowing destabilization and coagulation to occur.

Bridging is also a mechanism resulting from adsorption of hydrolysed metal ion coagulants due to tendencies of the hydrolysed species to polymerize thereby bridging adjacent particles on adsorption and promoting agglomeration, i.e. flocculation.

Precipitate enmeshment is another mechanism for removing colloidal particles from water and occurs under appropriate conditions of metal ion coagulant concentration (usually high) and pH. Formation of metal ion hydroxide precipitates serve to enmesh particulate material by a sweeping action which is usually independent of colloid concentration.

Bratby (1980) has adapted from Stumm and O'Melia (1968) a tabular representation of the characteristics of these mechanisms as given in Table 2.

James and Healy (1972) showed that the adsorption of hydrolysis products of metal ion coagulants occurs much more readily than the free aquo metal ion.

The effect of coagulant dosages is dependent on the predominant mechanism of destabilization. If for example, reduction of the thickness of the electrical double layer is the principal mechanism of destabilization the response to coagulation will occur as shown in Figure 2.

The critical coagulant concentration (CCC) is shown by the inflection point on the curve, and because the destabilization is solely charge neutralisation the CCC is independent of particle concentration and increasing coagulant dose has no effect. It also follows that the higher the valency of counter-ions (coagulant ion) the lower the CCC for destabilization.

Often, however, more than one destabilization mechanism is evident and is dependent on coagulant concentration, particle concentration and pH. An example of this more complex behaviour is shown in Figure 3.

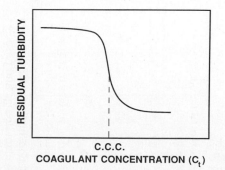

Figure 2.    Destabilization characteristics where repression of electrical double layer is the predominant mechanism (Stumm and O'Melia, 1968, reprinted by permission Uplands Press Ltd.).

With increasing initial coagulant concentration at constant pH, two regions of destabilization are observed. The first shown as $CCC_1$, represents destabilization by a charge reduction or neutralisation mechanism due to adsorption of counter-ion species or possibly by bridging of polymeric metal hydrolysis species. At higher initial coagulant dosages restabilization occurs at CSC, most likely because of charge reversal resulting from excess counter-ion adsorption.

If the initial coagulant dose is increased sufficiently destabilization again occurs at $CCC_2$ as shown, again by two possible mechanisms, the predominance of each being dependent on the colloid concentration. At high colloid concentration the predominant mechanism will be charge neutralisation or electrical double layer repression. Most likely at low colloid concentrations is particle enmeshment by the metal hydroxide precipitates in a sweep mechanism.

The effect of colloid concentration is represented graphically for a given pH value in Figure 4.

At low colloid concentrations (S1) there is little opportunity for colloid-colloid interaction and with coagulant dosage at a predicted stoichiometric level uneven adsorption occurs, resulting in some destabilization, but some particles remain stable. At a higher dosage the coagulant to colloid concentration ratio is high and restabilization due to excessive ion adsorption occurs. Therefore coagulant concentration must be increased to at least $C_1$ where metal hydroxide precipitates form and removal is by a sweep mechanism.

As colloid concentration is increased a point, A, is reached where contact opportunity is sufficient for hydrolysis species to become adsorbed and effect destabilization by charge reduction or bridging, thereby significantly reducing coagulant dosage. At concentrations such as S2 destabilization occurs over a relatively narrow range of coagulant dosage. Critical coagulant concentration ($CCC_1$) is much lower than for S1, but beyond a critical coagulant dosage (CSC) restabilization occurs and coagulant dose must be at least $CCC_2$ for metal hydroxide precipitation. Adsorption destabilization is the predominant mechanism in this region and as seen by the linearity of the curve the critical coagulant concentration is dependent on colloid concentration.

At higher colloid concentrations restabilization does not occur. The first effect is compression of the diffuse layer of charge and rather than extensive adsorption occurring the closer particle proximity will promote floc formation immediately following destabilization. As indicated when discussing Figure 2 colloid concentration has no effect

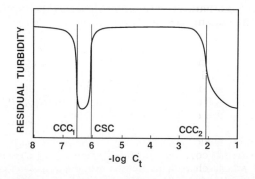

Figure 3.    Destabilization characteristics where adsorption of coagulant species to colloidal particles is operative. $CCC_1$, $CCC_2$ and CSC are the lower and upper critical coagulant concentrations and restabilization concentration of the coagulant respectively (Stumm and O'Melia, 1968, reprinted by permission Uplands Press Ltd.).

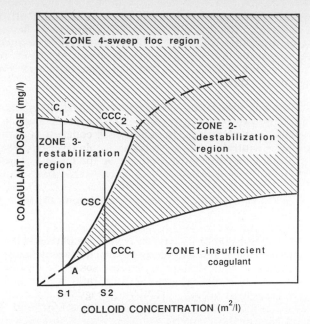

Figure 4.   Effect of colloid concentration, S, and coagulant dosage, $C_t$, on zones of destabilization and restabilization at a given pH value. Destabilization has occurred in the hatched area (Stumm and O'Melia, 1968).

on the coagulant dosage for destabilization.   At very high doses in this region destabilization can be effected by metal hydroxide precipitation.

Healy (1978), in discussing the variables that control the properties of aqueous interfaces, demonstrated the almost universal role of pH in controlling electrical properties at aqueous interfaces. It is important therefore that we consider the influence of pH on destabilization reactions with hydrolysable metal coagulants.

With the systems predominantly encountered in water treatment operations charge reversal of the colloidal dispersion is the feature mechanism of the destabilization of colloidal species.  Stumm and O'Melia (1962, 1968) showed by example that a critical coagulant concentration (CCC) and a critical stabilization concentration (CSC) are observed for a given pH and that plots of log CCC and log CSC against pH delineate the coagulation domain of the system.  It is also shown that CCC and CSC are dependent on colloid concentration.

James et al. (1977) demonstrated that for such systems a simple relationship exists between the zeta potential and the coagulation domains in a hydrolysable metal ion - colloid system.  On the basis of this relationship, Hutchison et al. (1985) examined the coagulation domains (zones of destabilization and restabilization) as a function of pH, colloid concentration, coagulant dose and coagulant type.

The electrophoretic mobility (which monitors the trends in zeta potential) results given in Figure 5 serve to illustrate this complex interrelationship.  Note that in water the turbidity (in this case colloidal silica, chosen to mimic quite accurately the vast majority of turbidity situations encountered in practice) is uniformly negative over a wide pH range, coagulation by pH change alone occurring only at a very low pH.  The addition of iron or aluminium coagulant produces a characteristic change in the surface or zeta potential vs pH curve, and at the higher doses includes a negative to positive change in potential and then with increases in pH a positive to negative change each passing through zero potential as previously demonstrated by Matijevic et al. (1971).

127

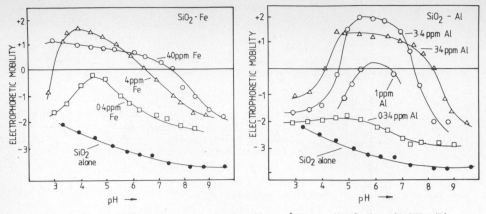

Figure 5.    Electrophoretic mobility ($\mu m\ s^{-1}/V\ cm^{-1}$) of silica (0.05 g/L) as a function of pH and iron and aluminium dose (in this case ferric and aluminium sulphate) (Hutchison et al., 1985).

Corresponding coagulation domains are shown in Figure 6 and at each dose coagulation is observed only in the pH range where the zeta potential is near or at zero. Of interest is the wider and deeper coagulation domain in the presence of iron coagulant and the absence of a stability or dispersion "window" as shown in the aluminium coagulation domain under identical conditions.

It is now possible to generalise delineation of coagulation domains which could be expected with Fe and Al coagulants in practice as shown in Figure 7.

The type of domain which could be encountered in practice for any given water will depend on:

•  The total available surface area of dispersed solids in the water, and

•  The total concentration of solutes or "solubles" which can enter into precipitation reactions with Fe or Al.

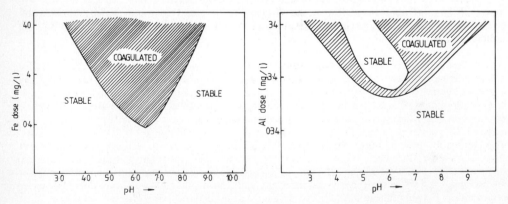

Figure 6.    Coagulation domains for iron and aluminium coagulation of colloidal silica (0.05 g/L) in terms of log (dose) vs pH. Systems were coagulated within the shaded region and dispersed elsewhere (Hutchison et al., 1985).

128

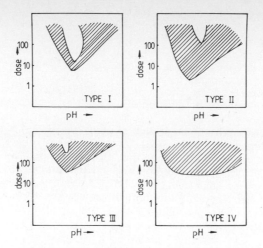

Figure 7.  Generalised log (dose) vs pH coagulation domains observed in Fe and Al coagulation processes for a wide range of waters. The systems are coagulated in the hatched area and dispersed elsewhere (Hutchison et al., 1985).

Consider the transition from Type I domain, with a narrow lower pH arm, large stability "window" and a sharp rise to a broader high pH arm, to the Type II domain with a small (sometimes nonexistent) stability "window" and broad lower and upper pH arms. For many waters the transition is observed when comparing the domains of Al and Fe coagulation on the same water.

Transition from Type I to Type III and from Type II to Type IV is observed when the turbidity or suspended/dispersed solids concentration increases or there is an increase in the total "solubles".  With increases in turbidity and/or precipitation, the domains broaden and move to the upper part of the dose range and again the much broader Type IV domain is more characteristic of Fe coagulation when compared with Al coagulation at higher turbidity/precipitate levels.  Similar coagulation domains were demonstrated by Vik et al. (1985) in the destabilization of organic colour with metal ion coagulants.


## INFLUENCE OF ANIONS

We have already discussed in a limited way the competitive nature of the OH$^-$ ligand during precipitation mechanisms between metal ion coagulants and hydrophilic colloids where the displacement of OH$^-$ ions by certain anions from the coordination sphere of metal coagulant ions occurs at lower pH values.

Therefore, (as first put forward by Thomas and Marion, 1946) if the anion is a strong coordinator with the metal coagulant ion and not readily displaced by OH$^-$ ions, the pH of optimum destabilization decreases rapidly with increasing anion concentration, but if it is readily displaced by OH$^-$ ion the optimum destabilization pH increases with a very basic anion and decreases with a weakly basic anion.  If the anion is a weak coordinator with the metal ion it exerts only slight effects of decreasing optimum pH values.

It then follows that nitrate ion has little influence on coagulation and chloride ion will, in relatively high concentrations, shift the optimum pH to the acid side. The effect of sulphate and phosphate ions have been examined and are illustrated by the results in Figure 8.

The alum only curve in both cases shows a typically narrow pH zone of optimum destabilization with sharp rises along the arms of the upper and lower pH zones of destabilization. The effect of the sulphate ion is to extend the pH zone of optimum destabilization towards the acid side. With increasing sulphate ion concentration the opportunity for mixed ligand complexes to form also becomes greater. The sulphate ion is strongly coordinated with metal ions but weakly basic and at lower pH, OH$^-$ is more readily displaced to form (Al)-(OH) complexes, whereas at pH greater than approximately 7, little effect results because minimal displacement of OH$^-$ occurs. At very low pH and high sulphate concentration (Al)-(SO$_4$) complexes predominate which have no significant absorptive capacity and therefore little effect on destabilization.

The effect of increasing the concentration of phosphate ion is to shift the pH zone of optimum destabilization to the acid side. At high (7 and above) pH values, the principal phosphate species are multicharged and strong coordinators with metal ions. These readily displace OH$^-$ from the coordination sphere, but render the complex less positive. At lower pH values singly charged phosphate species predominate and the availability of OH$^-$ is reduced, therefore to retain the same overall charge characteristics as at the optimum pH value with phosphate absent, the optimum destabilization pH in the presence of phosphate is lower.

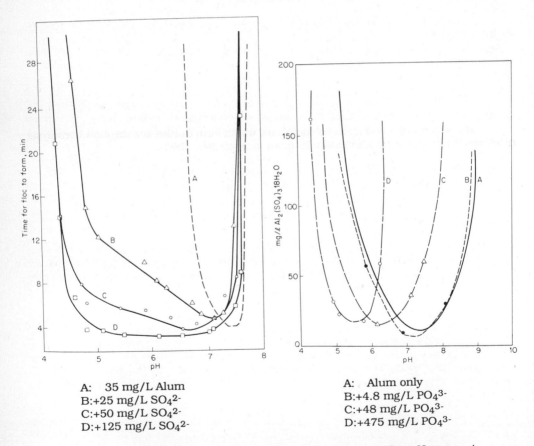

A:  35 mg/L Alum
B:+25 mg/L SO$_4^{2-}$
C:+50 mg/L SO$_4^{2-}$
D:+125 mg/L SO$_4^{2-}$

A:  Alum only
B:+4.8 mg/L PO$_4^{3-}$
C:+48 mg/L PO$_4^{3-}$
D:+475 mg/L PO$_4^{3-}$

Figure 8.   Effect of sulphate and phosphate ions on the pH zone of coagulation. Points on the curves for phosphate represent alum dose for a 50% turbidity reduction (From Packham, 1963 and Black et al., 1933 respectively. Reprinted by permission McGraw-Hill Book Co.).

## METAL ION COAGULANTS

Two broad categories used in Australia are those metal ion coagulants based on iron and on aluminium. Iron coagulants include ferric sulphate e.g. the Tioxide product known as FerriClear (trademark of Tioxide Australia Pty. Ltd.), ferrous sulphate and ferric chloride. The aluminium coagulants include aluminium sulphate, aluminium chloride, polyaluminium chloride (or PAC) and sodium aluminate.

Cost and effectiveness are the two main reasons for choice of coagulant and the other factor pertaining to the two categories of coagulants is their ability to form multicharged polynuclear complexes with enhanced adsorption properties in aqueous systems. Ferric and aluminium salts in solution form coordination compounds $[Fe(H_2O)_6]^{3+}$ and $[Al(H_2O)_6]^{3+}$. During destabilization reactions with hydrophobic colloids there will be stepwise substitution of water molecules by other ligands and ions, principally $OH^-$. Hence the importance of pH as a controlling factor.

The donor capacity of these ligands when coordinated to a metal ion is generally available for coordination with another metal ion and will act as a bridge. This is exhibited by iron and aluminium hydroxo complexes in their tendencies to form polynuclear complexes, illustrated by the simple reaction to form a binuclear iron complex.

$$2[Fe(H_2O)_5OH]^{2+} \longrightarrow [Fe_2(H_2O)_8(OH)_2]^{4+} + 2H_2O$$

The structure of which is believed to show bridging of the two Fe ions by two hydroxo bridges:

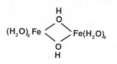

The species of metal hydroxo complexes which form during substitution hydrolysis reactions is illustrated for aluminium (Stumm and Morgan, 1962):

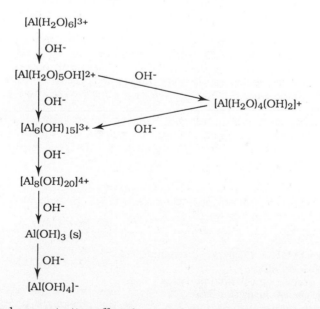

pH and concentration affect the rate of transformation from one species to another and during these reactions a number of polynuclear complexes form with the extent of polymerization generally becoming greater as the charge per metal ion on the hydroxo complexes decreases.

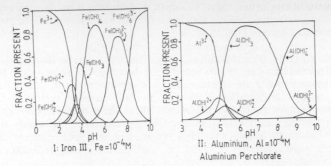

Figure 9.   Distribution of species for iron and aluminium during destabilization (Singley and Sullivan, 1968, 1969).


The results from Singley and Sullivan (1968, 1969) showing the distribution of species for iron and aluminium hydroxo complexes are illustrated in Figure 9.

A note of interest is the predominance of $Fe(OH)_3$ at pH 4.0, which is consistent with conditions for organic colour removal by precipitation. The prediction for aluminium reactions is not so clear because of the greater tendency for formation of polynuclear species particularly at low ($10^{-5}$ to $10^{-4}$ M) concentrations. This is partly responsible for the narrow pH range of optimum destabilization reactions compared with iron salts. It is noted, however, that greater polymerization enhances adsorption destabilization reactions.


## FLOCCULATION

We have previously discussed the reactions and mechanisms which destabilize the components we wish to remove in water treatment processes. Following destabilization the random collision of particles due to Brownian motion results in formation of "pin" floc. This process is called *Perikinetic* flocculation and is completed within a few seconds because of a limiting floc size beyond which Brownian motion has little effect.

This process occurs in what is usually known as the "flash-mixing" stage of the process, so it can be seen that the flash-mixer, necessary for efficient and intimate coagulant contact, also performs perhaps the most important operation in water treatment. The few seconds mentioned for completion of this process is really determined by the reaction times for the various steps involved as summarised in Table 3.

Careful consideration of flash-mixer choice relative to the dominant destabilization reactions determined, should be given. Plug flow mixers are ideally suited where adsorption destabilization reactions dominate, because retention time for all components is the same.


Table 3.   Approximate Times (Seconds) for Various Reactions Involved in the Destabilization Process (Bratby, 1980)

| | |
|---|---|
| Mononuclear complex ($AlOH^{2+}$, $FeOH^{2+}$) formation | $10^{-1}$ |
| Polynuclear complexes formation | $10^{-2}$ - 1 |
| Adsorption of above species on colloidal particles | $10^{-10}$ |
| Changes in electrical double layer structure | $10^{-8}$ |
| Brownian collision | $10^{-2}$-$10^{2}$ |
| Charge changes at particle surface | $10^{-6}$-$10^{4}$ |

Where mechanisms other than adsorption destabilization (i.e. sweep coagulation, sludge conditioning, polyelectrolytes) are dominant an extensive guide to process choice is provided by Vrale and Jordan (1971).

The next stage of flocculation is known as *Orthokinetic* flocculation and in this process further particle contact is promoted by inducement of velocity gradients either by physical obstructions in the flow path (baffles, filter media as in direct filtration) or by mechanical or hydraulic agitation. The differential particle velocities thereby create opportunity for collision and agglomeration. The time and velocity gradient factors determine both the rate and extent of particle aggregation. Obviously a high velocity gradient (high shear) will induce fast aggregation but will limit particle or floc size. It follows that for any given velocity gradient there will be a limiting flocculation time beyond which floc size will not increase.

A third stage of flocculation has been investigated by Mitsuo (1977) and is termed *Mechanical Syneresis*. This is defined as the process of exuding water from loose, bulky floc to further concentrate the floc particles. The increases in density then enable more efficient separation of floc and water. The process seems most suited to high turbidity water with polyelectrolyte dosing to increase floc particle strength.


## CONCLUSIONS

The importance of surface and near surface reactions in the make up of our natural waters and in shaping our water treatment processes has been discussed. The role of destabilization reactions has been emphasised along with the transition from stable systems through physico-chemical destabilizations to achieve a stable product of those components which we wish to efficiently separate from the water in the water treatment process.


## REFERENCES

Black, A.P. and Christman, R.F., 1963. Characteristics of coloured surface waters, *J. Amer. Waterworks Assoc.*, 55:753.

Black, A.P. and Christman, R.F., 1963. Chemical characteristics of fulvic acids, *J. Amer. Waterworks Assoc.*, 55:897.

Black, A.P., Rice, O. and Bartow, E., 1933. Formation of floc by aluminium sulphate, *Ind. Eng. Chem.*, 25:811.

Black, A.P., Singley, J.E., Whittle, G.P. and Maulding, J.S., 1963. Stoichiometry of the coagulation of colour causing compounds with ferric sulphate, *J. Amer. Waterworks Assoc.*, 55:1347.

Bratby, J., 1980. "Coagulation and flocculation", Uplands Press.

Bursill, D.B., Hine, P.T. and Morran, J.Y., 1985. The effect of natural organics on water treatment processes, *in:* "Proc. 11th Fed. Conv. Aust. Water & Wastewater Assoc.", Melbourne, p 197.

Donnan, M.B., Healy, T.W. and Nelson P.F., 1981. An electrokinetic study of alum coagulation and polymer flocculation of cellulose pulp fines, *Coll. Surf.*, 2:133.

Healy, T.W., 1978. Fundamentals of adsorption from aqueous solution, *in:* "Adsorption from Aqueous Solution", D.E. Yates, ed., RACI Colloid and Surface Chemistry Division, Melbourne, p 73.

Hine, P.T. and Bursill, D.B., 1984. The characterization of organics in water, *in:* "Aust. Water Res. Council: Technical Paper No. 86, AGPS.

Hutchison, P.R., Freeman, P. and Healy T.W., 1985. Principles and applications of iron coagulants in water treatment, *in:* "Proc. 11th Fed. Conv. Aust. Water & Wastewater Assoc.", Melbourne, p 222.

James, R.O. and Healy, T.W., 1972. Adsorption of hydrolysable metal ions at the oxide - water interface, *J. Coll. Interface Sci.*, 40:42.

James, R.O., Wiese, G.R. and Healy, T.W., 1977. Charge reversal coagulation of colloidal dispersions of hydrolysable metal ions, *J. Coll. Interface Sci.*, 59:381.

La Mer, V.K. and Healy, T.W., 1963. The role of filtration in investigating flocculation and redispersion of colloid dispersions, *J. Phys. Chem.*, 67:2417.

Lyklema, J., 1981. Fundamentals of electrical double layers in colloidal systems, *in:* "Colloidal Dispersions", J. Goodwin Ed., Royal Soc. Chem., London, p.47.

Matijevic, E., Mangravite, F.J. and Cassell, E.A., 1971. Stability of colloidal silica IV. The silica-alumina system, *J. Coll. Interface Sci.*, 35:560.

Mitsuo Yusa, 1977. Mechanisms of pelleting flocculation, *Int. J. Min. Process*, 4:293.

Narkis, N. and Rebhun M., 1977. Stoichiometric relationship between humic and fulvic acids and flocculants, *J. Amer. Waterworks Assoc.*, 69:325.

O'Melia, C.R. and Stumm, W., 1967. Aggregation of silica dispersions by iron (III), *J. Coll. Interface Sci.*, 23:437.

Overbeek, J. Th. G., 1973. Recent developments in the understanding of colloid stability, *J. Coll. Interface Sci.*, 58:408.

Packham, R.F., 1963. The coagulation process - a review of some recent investigations, *in:* "Proc. Soc. Water. Treat. Exam.", 12:15.

Priesing, C.P., 1962. A Theory of Coagulation Useful for Design, *Ind. Eng. Chem.*, 54:838.

Singley, J.E. and Sullivan, J.H., 1969. Reactions of metal ions in dilute solutions: recalculation of hydrolysis of iron(III) data, *J. Amer. Waterworks Assoc.*, 61:190.

Sinsabaugh, R.L., Hoehn, R.C., Knocke, W.R. and Linkins, A.E., 1986. Removal of dissolved organic carbon by coagulation with iron sulphate, *J. Amer. Waterworks Assoc.*, 78:74.

Stumm, W. and Morgan, J.J., 1962. Chemical aspects of coagulation, *J. Amer. Waterworks Assoc.*, 54:971.

Stumm, W. and O'Melia, C.R., 1968. Stoichiometry of coagulation, *J. Amer. Waterworks Assoc.*, 60:514.

Sullivan, J.H. and Singley, J.E., 1968. Reactions of metal ions in dilute aqueous solution: hydrolysis of aluminium, *J. Amer. Waterworks Assoc.*, 60:1280.

Thomas, A.W. and Marion, S.P., 1946. effect of diverse ions on the pH of maximum precipitation of aluminium hydroxide, *J. Coll. Sci.*, 1:221.

Vik, E.A., Carlson, D.A., Eikum, A.S. and Gjessing, E.T., 1985. Removing aquatic humus from Norwegian Lakes, *J. Amer. Waterworks Assoc.*, 77:58.

Vrale, L. and Jordan, R.M., 1971. Rapid mixing in water treatment, *J. Amer. Waterworks Assoc.*, 63:52.

# ENHANCED BIOLOGICAL REMOVAL OF

# PHOSPHORUS FROM WASTEWATER

Annabelle Duncan[1], Ronald C. Bayly[1],
John W. May[1], George Vasiliadis[1] and
William G. C. Raper[2]

[1]Department of Microbiology
Monash University
Clayton, Victoria

[2]CSIRO Division of Chemicals and Polymers
Clayton, Victoria

## INTRODUCTION

Eutrophication, especially of inland waters, has emerged as one of the more severe pollution problems in arid nations such as Australia. This is caused mainly by the entry of excessive amounts of nitrogen (N) and phosphorus (P) into the waters. Removal of compounds of N and P from effluents is an important factor in preventing eutrophication.

Biological methods for the removal of nitrogen from sewage via nitrification and denitrification are reasonably well understood and are utilized in many sewage treatment processes. Biological phosphorus removal is not as well understood and, until recently, has received little attention. This is due in part to the prevalence of chemical methods for phosphorus removal. Addition of ferric or aluminium salts or of lime to effluents cause precipitation of phosphates which can then be removed from wastewater with sludge. But there are many disadvantages in chemical methods, including the increased volumes of sludge that have to be disposed of, undesirably high or low pH of the effluent and very high running costs for chemicals and maintenance (Shoda et al.,1980). In fact, in recent years, costs of chemical treatment have increased dramatically and alternatives have been sought. Conventional biological sewage treatment processes remove 50% or less of the phosphorus in sewage. Removals of >90% are usually required to reach acceptable effluent concentrations of <1 mg/L. In these biological processes, N and P are incorporated into the biomass produced during oxidation of carbonaceous compounds. The amount of P removed will depend upon the requirements for growth of the cells, while the percentage removed will depend on the ratio of phosphorus to biodegradable carbon present in the influent. Under normal operating conditions the biomass removes about 1 part phosphorus per 100 parts of organic carbon (Odum, 1959). This is much less than the ratio normally found in domestic sewage.

In the 1960's it was noted in some activated sludge plants in India and in America, that removal of phosphorus took place in excess of that needed for the normal metabolic requirements of the microbiota (Marais et al.,1983). Closer examination of these plants revealed that oxygen input had been restricted at the start of the activated sludge process, causing anaerobic conditions to develop. A phosphorus profile across one of these plants showed that phosphorus was actually released into solution in the

*Surface and Colloid Chemistry in Natural Waters and Water Treatment*
Edited by R. Beckett, Plenum Press, New York, 1990

anaerobic zone, but that excess uptake occurred in the aerated areas, i.e. more phosphorus accumulated than was initially released.

There was some controversy as to whether the mechanism for this phenomenon was chemical or biological. Menar and Jenkins (1970) claimed that phosphorus was removed as a result of chemical precipitation. They suggested that the aeration process stripped metabolically produced $CO_2$ from the sludge, causing an increase in pH and inducing precipitation of calcium phosphate which was then removed with waste sludge. This model does not explain the requirement for the anaerobic stage. Other workers investigated the process further (Carberry and Tenney, 1973) but have found no correlation between pH change and phosphorus removal. No significant increase in pH was observed during the time that maximum phosphorus removal occurred.

There was also little calcium removal from the liquid phase during this time. Furthermore, extraction experiments showed that most accumulated phosphorus was associated with cells and not with extracellular sludge solids and, if biological processes were "poisoned" with 2,4,-dinitrophenol (an uncoupler of oxidative phosphorylation) no phosphate uptake was observed.

This work together with results of several other workers (Barnard, 1976; Hoffman and Marais, 1977; Rensink et al., 1981) have been taken as proof that phosphorus removal from activated sludge is a biological process. This removal of more phosphorus than is required by the normal metabolic demands of the microbiota is referred to as enhanced biological phosphorus removal.

These observations caused considerable interest, especially in South Africa, where water shortage has led to the construction of dams on many of the rivers and where eutrophication is a major problem. Several sewage treatment plants were constructed in South Africa that were designed specifically to remove both nitrogen and phosphorus by biological processes (Nicholls, 1975).

The basic design of these plants is shown in Figure 1. Figure 1a is a conventional activated sludge plant. This process depends on the formation of a gelatinous floc (the activated sludge), of bacteria and protozoa which flocculates particulate and colloidal materials. These are then degraded together with soluble biodegradable substances by the microorganisms in the floc. Activated sludge therefore has the ability to remove colloidal and dissolved organic material from the liquid phase to the solid phase, which will settle out in the clarifier (which is not aerated) (Nicholls et al.,1984). The effectiveness of the process depends on a return of a portion of the settled sludge, with its associated microorganisms, to the aerated zone, to inoculate the incoming sewage (Tebbut, 1977).

The process was modified for phosphorus and nitrogen removal by including two further zones, the anoxic zone which encourages reduction of nitrate to gaseous nitrogen (denitrification) and the anaerobic zone where phosphorus is released, thus priming the cells in some way for uptake in the aerobic zone (Figure 1b). As used in this context, anaerobic means lack of oxygen and oxidized nitrogen compounds such as nitrate and nitrite and anoxic means a lack of oxygen, but nitrate and nitrite may be present.

Some of these plants worked well but others showed poor phosphorus removal. The major problem was traced to a failure to achieve strictly anaerobic conditions due to the presence of nitrates in the "anaerobic" zone. High concentrations of nitrate prevented phosphate release and without release, no subsequent uptake occurred in the aerobic zone. Plants with dilute feed (i.e. low carbon content) also exhibited poor nutrient removal. In these cases insufficient carbon was available in the feed to allow complete denitrification so that the concentration of nitrate in the return feed was high (Nicholls et al., 1984). Two solutions were found:

1.    Discharging the return sludge into the anoxic reactor, where denitrification removes nitrate, and then recycling sludge from this reactor to the anaerobic zone.

2.    Increasing the carbon content of the feed by adding extra high carbon effluent. It

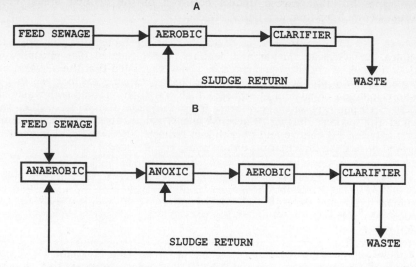

Figure 1.   Configuration of activated sludge plants (A) conventional plant (B) plant modified for N and P removal (anaerobic - no oxygen, nitrate or nitrite present; anoxic - no oxygen present; aerobic - oxygenated).

has been claimed that enhanced phosphorus removal will not occur if the C:N ratio of the sewage is <14:1. But good removal efficiencies have been noted at lower ratios if volatile fatty acids e.g. acetate are present (Barnard, 1984). The addition of acid digested sludge to dilute feed has been found to be particularly successful.

However, problems were still encountered in some plants and the actual mechanism of phosphorus removal from activated sludge remains controversial. There is now much interest in understanding the microbial processes involved, so that better control of treatment plants will be possible.

## BIOCHEMISTRY OF PHOSPHORUS UPTAKE BY MICROORGANISMS

Phosphorus enters microbial cells as orthophosphate, which is utilized to form adenosine triphosphate (ATP) (the main energy supply of the cell) and much of this ATP is used to produce nucleic acids. This inhibits accumulation of polyphosphates.

Under some circumstances, however, phosphorus accumulates as polyphosphate (or volutin) granules which can be easily visualised microscopically within the cell. If growth, and consequently nucleic acid synthesis stop because of limitation of nutrients such as sulphur or nitrogen, competition for ATP is relieved and polyphosphate can accumulate. This is known as "luxury uptake".

In cells that have been subjected to phosphate starvation, the polyphosphate uptake system becomes operative. If these cells are then supplied with phosphate it will be rapidly taken up and polyphosphate synthesized. This phenomenon is known as polyphosphate "overplus" or "overcompensation" (Osborn and Nicholls, 1978).

Some researchers claim that some microorganisms must be capable of polyphosphate accumulation under balanced growth conditions to account for uptake in activated sludge (Fuhs and Chen, 1975).

Polyphosphate is thought to have two major functions:

1.    It can supply the cell with energy.    When polyphosphate is degraded to orthophosphate, energy is released.

2.    It can supply the cell with phosphorus for metabolic purposes if the external supply is low.

Pilot plant experiments have revealed that low effluent phosphorus concentrations were only possible if polyphosphate-containing bacteria were present (Osborn and Nicholls, 1978). The organisms which accumulated polyphosphate were all strict aerobes and yet it is known that polyphosphate removal occurs only in plants with an anaerobic zone.

Most workers have implicated *Acinetobacter* as the major genus involved in enhanced phosphorus removal (Fuhs and Chen, 1975; Nicholls et al., 1984 ). While some groups have found that *Acinetobacter* spp. comprise the greatest proportion of the polyphosphate accumulators, most find similar "counts" of these organisms in successful and unsuccessful plants (Lotter et al., 1986). There are several possible reasons for this:

1.    *Acinetobacter* spp. have a tendency to clump together when they accumulate polyphosphate. This makes accurate "counting" by conventional cultural methods difficult.    For example, in work carried out in our laboratory we found approximately 10-fold fewer *Acinetobacter* in sludge from a pilot plant exhibiting enhanced phosphorus uptake than there had been in the same plant at start-up before enhanced P uptake occurred. However, microscopic observations revealed the phosphate accumulating bacteria were present as single, well dispersed cells at start-up but formed compact clusters later. Each of these clusters would have been counted as a single organism using cultural methods although we estimated that each could have contained at least 1000 cells (Vasiliadis et al., 1987). Tetrazolium reduction assays suggest that all the cells in a cluster may be viable. This illustrates one of the problems of depending upon cultural methods as most early studies did.

2.    Conditions in successful plants may exert positive selective pressure for particular subgroups of *Acinetobacter* that cannot be distinguished from other subgroups by conventional biochemical tests.

3.    *Acinetobacter* may always be present but may accumulate polyphosphate only under very specific conditions.

Many workers have also observed that the bacterial storage compound, poly-β-hydroxybutyrate (PHB), accumulates in cells during the anaerobic phase and disappears during the aerobic phase, whereas polyphosphate does the opposite (Fuhs and Chen, 1975; Nicholls and Osborn, 1979).

Any model of enhanced biological phosphate removal must therefore take into account the following points:

1.    The plant must include an anaerobic zone where phosphate release must occur. Release is increased by the presence of acetate.

2.    Nitrate will prevent phosphate release and so no subsequent accumulation is possible.    Nitrate inhibition is relieved by high concentrations of carbonaceous compounds, especially acetate (Iwema and Meunier, 1985).

3.    The organisms which are capable of polyphosphate accumulation are all strict aerobes and accumulation only occurs under aerobic conditions.

4.    In plants showing enhanced phosphorus removal the microbiota accumulates PHB during the anaerobic stage but this is apparently utilized during the aerobic phase.

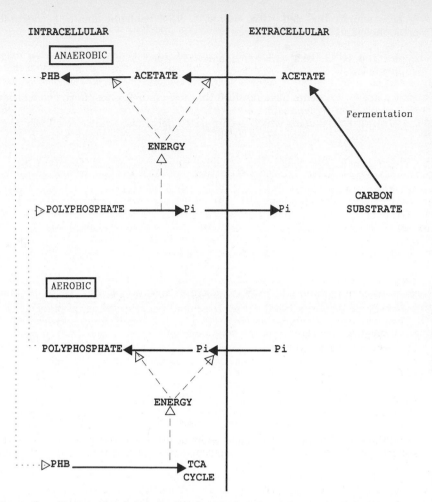

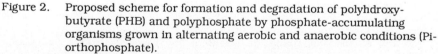

Figure 2.    Proposed scheme for formation and degradation of polyhdroxy-
butyrate (PHB) and polyphosphate by phosphate-accumulating
organisms grown in alternating aerobic and anaerobic conditions (Pi-
orthophosphate).

Several models have been proposed to explain biological phosphorus removal from
activated sludge (Buchan, 1983; Marais et al., 1983; Comeau et al., 1986); most are
based on the premise that under anaerobic conditions acetate is produced by
fermentation by other bacteria, possibly *Aeromonas*.    The acetate is taken up by
polyphosphate-accumulating bacteria (called Bio-P bacteria from here on).   Because the
Bio-P bacteria are obligate aerobes i.e. they require oxygen as the terminal electron
acceptor for respiration, they are incapable of regenerating their major energy  supply
(ATP) under anaerobic conditions. Instead, they use polyphosphate as their energy source
for maintenance of membrane function and uptake of acetate, releasing orthophosphate
into the liquid phase.    The accumulated acetate is converted to PHB, the storage
compound observed to build up during the anaerobic phase.   In this way, the cells are
depleted of polyphosphate.  Because they are aerobes, they are incapable of forming more
polyphosphate under anaerobic conditions.    They are thus effectively in a state of
phosphorus deficiency (despite the high concentration in the surrounding medium) and
so are derepressed for phosphate accumulation.    This is one mechanism whereby
polyphosphate accumulation could occur via the "overplus" mechanism, once oxygen is
resupplied.  The PHB accumulated in the cells during the anaerobic phase provides the
cells with a readily utilizable carbon source, so that when oxygen is resupplied, the cells

can multiply rapidly without competing with other aerobes for exogenous carbon sources and can accumulate polyphosphate by the "overplus" mechanism (Figure 2).

Controversy still exists over several stages of this composite model. For example, there is dispute over the mechanism of nitrate "poisoning". Some claim that if nitrate is present it can be used instead of oxygen as terminal electron acceptor in the regeneration of ATP and so polyphosphate is not broken down. There is thus no phosphorus deficiency and no "overplus" uptake. An alternative view is that if nitrate is present then separate nitrate-reducing organisms can compete with the Bio-P organisms for the acetate.

Some workers also claim that an anaerobic stage is unnecessary and that, providing acetate is supplied, Bio-P organisms can continue to accumulate phosphorus.

There is also dispute over the role of PHB. Some workers have found it consistently in the anaerobic phase, others have not. We can detect PHB in laboratory scale systems using sewage as an inoculum but not when we use pure cultures of Bio-P accumulators. Still other researchers claim that PHB synthesis from acetate is not possible under anaerobic conditions (Marais et al., 1983).

These points need to be resolved, as do the effects of temperature and influent composition, the importance of nitrate, acetate and of cations such as calcium, magnesium and potassium. Some workers claim that "overplus" accumulation cannot occur in the absence of calcium, others claim that the magnesium and potassium are important in balancing the charge of the polyphosphate.

## CONCLUSIONS

There is increasing interest in Australia in the process of enhanced biological phosphorus removal. Like South Africa, Australia is a very arid country where low water flows can cause severe problems with eutrophication, especially in inland waterways. For example, much of the summer flow of the Yarrowee Creek which flows from Ballarat into the Barwin River consists of effluent from a sewage treatment plant. High concentrations of nitrogen and phosphorus in this effluent will cause severe problems, but to remove phosphorus chemically would cost an estimated $200,000-300,000 per annum. This is for a community of only 50,000-70,000 people. Biological removal of phosphorus from the wastewater is promising but >90% removal cannot always be obtained in full-scale plant operation. None-the-less a pilot scale plant is at present being operated at Ballarat.

We are at present trying to elucidate the mechanisms of polyphosphate accumulation and to reconcile some of the conflicting theories. Greater understanding of the process will allow greater control of the activated sludge plants and will therefore result in reliable performance.

## ACKNOWLEDGMENTS

This work is part of an on-going collaborative research programme between the Microbiology Department, Monash University, Clayton, Victoria and the CSIRO, Division of Chemicals and Polymers, Clayton, Victoria. A. Duncan is supported by an Australian Water Research Council Postdoctoral Fellowship and G. Vasiliadis by the Australian Chemical Specialties Manufacturers Association and an extramural grant from CSIRO.

## REFERENCES

Barnard, J.L., 1976. A review of biological phosphorus removal in the activated sludge process, *Water SA*, 2:136-144.

Barnard, J.L., 1984. Activated primary tanks for phosphate removal, *Water SA*, 10:121-126.

Buchan, L., 1983. Possible biological mechanism of phosphorus removal. *Water Sci. Technol.*, 15:87-103.

Carberry, J.B. and Tenney, M.W., 1973. Luxury uptake of phosphate by activated sludge, *J. Water Poll. Control Fed.*, 45:2444-2462.

Comeau, Y., Hall, K.J., Hancock R.E. and Oldham, W.K., 1986. Biochemical model for enhanced biological phosphorus removal, *Water Res.*, 12:1511-1521.

Fuhs, G.W. and Chen, M., 1975. Microbiological basis of phosphate removal in the activated sludge process for the treatment of wastewater, *Microbial Ecol.*, 2:119-138.

Hoffmann, R. and Marais, G.V.R., 1977. "Phosphorus removal in the modified activated sludge process", Research Report W 22, Department of Civil Engineering, University of Cape Town.

Iwema, A. and Meunier, A., 1985. Influence of nitrate on acetic acid induced biological phosphate removal, *Water Sci. and Technol.*, 17:289-294.

Lotter, L.H., Wentzel, M.C., Lowenthal, R.C., Ekama G.A. and Marais, G.V.R., 1986. A study of selected characteristics of *Acinetobacter* spp. isolated from activated sludge in anaerobic/anoxic/and aerobic systems, *Water SA*, 12:203-207.

Marais, G.V.R., Loewenthal, R.E. and Siebritz, I.P., 1983. Observations supporting phosphate removal by biological excess uptake - a review, *Water Sci. Technol.*, 15:15-41.

Menar, A.B. and Jenkins, D., 1970. Fate of phosphorus in wastewater treatment processes: enhanced removal of phosphate by activated sludge, *Environ. Sci. Technol.*, 4:1115-1121.

Nicholls, H.A., 1975. Full scale experimentation on the new Johannesburg extended aeration plants, *Water SA*, 1:121-131.

Nicholls, H.A. and Osborn, D.W., 1979. Bacterial stress: prerequisite for biological removal of phosphorus. *J. Water Poll. Control Fed.*, 51:557-569.

Nicholls, H.A., Osborn, D.W., Buchan, L., Melmed, L.N. and A.R. Pitman, 1984. Biological removal of phosphorus and nitrogen from waste water, *in:* "Continuous Culture: Biotechnology, Medicine and the Environment", A.C.R. Dean, D.C, Elwood and C.G.T. Evans, eds.. Ellis Horwood, Chichester, pp 185-204.

Odum, E.P., 1959. "Fundamentals of Ecology", W.B. Saunders, Philadelphia, p 79.

Osborn, D.W. and Nicholls, H.A., 1978. Optimization of the activated sludge process for the biological removal of phosphorus, *Progress in Water Technol.*, 10:261-277.

Rensink, J.H., Donker H.J.G.W. and De Vries, H.P., 1981. Biological P removal in domestic wastewater by the activated sludge process, *in:* "Proc. 5th European Sewage and Refuse Symposium", Munich, June, pp 487-502.

Shoda, M., Ohsumi, Y. and Udaka, S., 1980. Screening for high phosphate accumulating bacteria, *Agricultural and Biological Chem.*, 44:319-324.

Tebbut, T.H.Y., 1977. "Principles of Water Quality Control", University of Birmingham, England, pp 126-131.

Vasiliadis, G., Bayly, R.C., May J.W. and Raper, W.G.C., 1987. Changes in concentrations of polyphosphate-containing bacteria and *Acinetobacter* spp. during establishment of enhanced P removal, *IAWPRC Newsletter*, 1.

## ADDITIONAL READING

Brodisch, K.E.U. and Joyner, S.J., 1983. The role of microorganisms other than *Acinetobacter* in biological phosphate removal in activated sludge processes, *Water Sci. Technol.*, 15:117-125.

Comeau, Y., Rabinowitz, B., Hall, K.J. and Oldham, W.K., 1987. Phosphate release and uptake in enhanced biological phosphorus removal from wastewater, *J. Water Poll. Control Fed.*, 59:707-715.

Fukase, T., Shibata, M. and Minyaji, Y., 1982. Studies on the mechanism of biological phosphorus removal, *Japan J. Water Poll. Res.*, 5:309-317.

Gerber, A., Mostert, E.S., Winter C.T. and de Villiers, R.H., 1986. Interactions between phosphate, nitrate and organic substrate in biological nutrient removal processes. *Water Sci. Technol.*, 19:183-194.

Levin, G.V., Topol, G.J. and Tarnay, A.G., 1975. Operation of full-scale biological phosphorus removal plant, *J. Water Poll. Control Fed.*, 47:577-590.

Ramadori, R., 1987. ed., "Biological Phosphate Removal from Wastewaters", Pergamon Press, Oxford.

Wentzel, M.C., Lotter, L.H., Loewenthal, R.E. and Marais, G.V.R., 1986. Metabolic behaviour of *Acinetobacter* spp. in enhanced biological phosphorus removal - a biochemical model, *Water SA*, 12:209-223.

# SOME EFFECTS OF DAM DESTRATIFICATION

## UPON MANGANESE SPECIATION

Barry Chiswell and Mazlin Bin Mokhtar

Department of Chemistry
University of Queensland
St. Lucia, Queensland

## INTRODUCTION

Although the current World Health Organization recommended limit for manganese concentration in drinking water is set to 100 µg/L, water authorities in Queensland and in other parts of Australia find this figure well in excess of the level of manganese they will allow in their treated water. Regular output of potable water with 100 µg/L of Mn into a reticulation system is guarantied to lead to black pipe deposits and washing stains - the dirty water phenomenon. Thus the Gold Coast City Council in Southeast Queensland is currently keeping the allowable Mn level in treated water at less than 10 µg/L.

To treat a storage dam as a homogeneous body of water, all of which may be drawn upon at any time as a source of potable water, has long been recognized as a recipe for major problems (Hutchinson, 1957). A primary classification of a dam identifies four horizontally stratified regions which are of importance in the study of manganese and iron speciation. These are:

- the epilimnion
- the hypolimnion
- the metalimnion or thermocline
- and the sediment.

The basis for the stratification of the first three regions given above, was claimed to be due to temperature differences in the water body leading to variation of water density with depth (Allaby, 1977). Although during winter and early spring, natural water bodies remain isothermal over their total depth, the onset of warmer weather in the late spring is accompanied by heat addition to the upper water level at a rate which exceeds the ability of the lower layers to absorb such heat. As the water in the upper zone becomes less dense than that below it, the resistance to mixing increases. Wetzel (1975, p 69) notes that a temperature difference of only two or three degrees between the upper epilimnic and the lower hypolimnic zones will be enough to prevent complete circulation of the entire water column.

Whereas the period of stratification of dams and lakes in tropical and subtropical areas of Australia occurs over the period early spring (September) until the end of autumn (May), lakes in the temperate regions of the Northern Hemisphere have much shorter periods of stratification, viz. from late spring (May) through to early autumn (September) (McMahon, 1969; Nyffler, 1983; Chiswell and Rauchle, 1986). Dams in the cooler regions of Australia also appear to be stratified over a shorter period of the year (Welsh, 1984,

*Surface and Colloid Chemistry in Natural Waters and Water Treatment*
Edited by R. Beckett, Plenum Press, New York, 1990

143

p 58), although it would appear that they are still stratified for up to eight months and thus for a longer time than the four to five months for lakes in Northern Europe or America.

Nonetheless, as discussed fully by Wetzel (1975), the results of the annual stratification of large bodies of freshwater are much the same around the world, with marked similarities in the behaviour of manganese in the water column regardless of the geographical location of the water body (Delphino and Lee, 1968; Davison and Woof, 1984; Davison et al., 1982; Burns, 1981). The length of the period of stratification in warmer climates may create problems for water treatment personnel for a longer period of the year.

We are currently studying manganese speciation in all regions of the dam but the work to be described here will not mention sediment speciation. Details of the speciation of manganese in large water bodies has been previously discussed (Chiswell and Rauchle, 1987). The aim of the work described here was to study various dam parameters, in particular dissolved oxygen and manganese speciation, both before and after artificial destratification of the dam by forced aeration.

## EXPERIMENTAL

Both dams dealt with in this work (Hinze Dam and North Pine Dam) are close to Brisbane (within 70 km) and supply water to urban areas in the south east corner of the state of Queensland, Australia.

Sampling of dam water was undertaken using a double-inlet water sampler, while temperature and dissolved oxygen measurements were obtained using a submersible oxygen/temperature probe.

Water samples were stored in polythene bottles which had been previously acid washed and then thoroughly rinsed in either distilled or "Milli-Q" water. Previous work had shown that either water could be used with no variation in the manganese concentrations (both soluble and insoluble) in the sample.

Filtration of samples was through either 0.45 μm or 0.01 μm cellulose acetate membranes (Sartorius). The latter filtrations were undertaken using an ultrafiltration unit.

Soluble manganese was determined on HCl-acidified filtrates with a Varian AA875 spectrometer, using the 279.5 nm emission line and an air-acetylene flame. Insoluble manganese was determined by a similar process after dissolution of the membrane and residue in concentrated HCl and dilution of the solution to the volume of the original sample from which it came.

## RESULTS AND DISCUSSION

While the thermocline is well established in Queensland dams from about October through May, with large concentrations of 0.45 μm filterable manganese in the hypolimnion, during a brief winter period it disappears and the whole body of the dam water becomes more or less homogeneous, taking on many of the characteristics of the summer epilimnic region. Thus, in this winter period, the dam has high dissolved oxygen and low dissolved manganese throughout the vertical profile.

In the hope of restoring homogeneous conditions during summer, engineers have attempted to artificially destratify the dam by blowing air through the dam waters. An aeration system is sunk to near the dam base and air pumped through the water body (Hutchinson, 1957). The success of artificial destratification depends on the design of suitable equipment for individual dam situations.

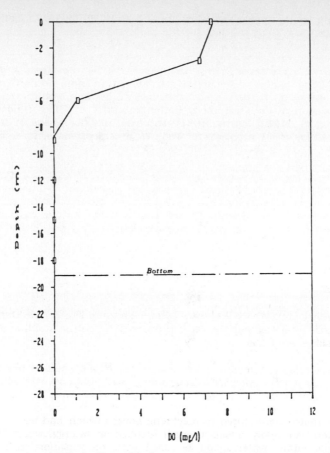

DO (mg/l)

Figure 1. Typical DO depth profile for a stratified dam (Hinze Dam, 22-1-85).

Results (Brown et al., 1982) published on the destratification of Lake Morris near Cairns (Northern Queensland, Australia) indicate that iron and manganese concentrations can be significantly reduced, for example from concentrations of manganese at the eight metre depth of approximately 1000 µg/L down to approximately 200 µg/L. However, there are some worrying features associated with a steady build-up of manganese in the water column during the years following the first clear initial success of the project. This is not unexpected, as removal of Mn from the water to the sedimental base will, in the long run, yield more sedimental manganese able to be diagenetically remobilized. Therefore one might expect the program to be a delaying mechanism rather than a long-term solution. It should also be pointed out that the total manganese values in Lake Morris are very high, even after destratification, thus the relative success of the process is not of great comfort if one wishes to reduce Mn down to below 50 µg/L.

The movement of the thermocline (based on temperature change) has been used to measure the onset of summer stratification, however, descending across the thermocline region one also has a lowering of pH and dissolved oxygen (DO) (Figure 1) and a marked increase in manganese concentration.

In our studies upon destratification and manganese speciation we prefer to see the metalimnion in terms of an oxycline rather than a thermocline. The oxycline, being the region over which there is a sudden change in DO values, may often coincide with the thermocline but is usually much more clearly defined.

To demonstrate that Hinze Dam had a typical stratification pattern to similar dams in the south east region of Queensland, some data on a dam some 70 km north

145

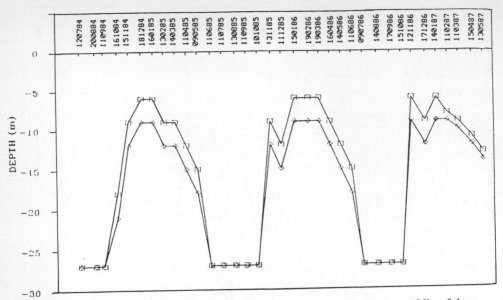

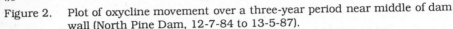

Figure 2.   Plot of oxycline movement over a three-year period near middle of dam wall (North Pine Dam, 12-7-84 to 13-5-87).

(North Pine Dam) were obtained.   Figure 2 is a plot of the position of the oxycline for North Pine Dam near Brisbane over a three year period. The points on the time axis are four weeks apart, while the top line shows the lowest depth with high DO, and the bottom line the highest depth with low or zero DO. These lines are usually about three metres apart and the oxycline is in this region.   When the two lines merge, the dam is fully oxygenated to the base - this occurs in winter. The pattern is remarkably regular and to check its validity a similar plot was done, but with all the time axis points shifted by two weeks (Figure 3). It should be noted that this dam was not being artificially aerated during this study.

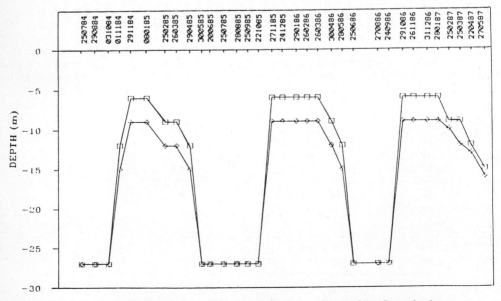

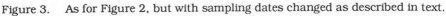

Figure 3.   As for Figure 2, but with sampling dates changed as described in text.

146

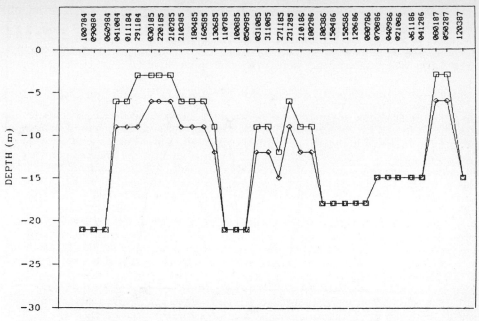

Figure 4.   Plot of oxycline movement over a three-year period at lower raw water
intake (Hinze Dam, 10-7-84 to 12-3-87).

The results are remarkably similar to oxycline plots for the Hinze Dam before
artificial destratification was commenced in October 1985. Figures 4 and 5 show the
change in oxycline profiles following October 1985, with full removal of the oxycline by
March 1986.    These two profiles were developed by plotting results obtained
approximately four weeks apart with the points in Figure 5 all two weeks later than those
in Figure 4.

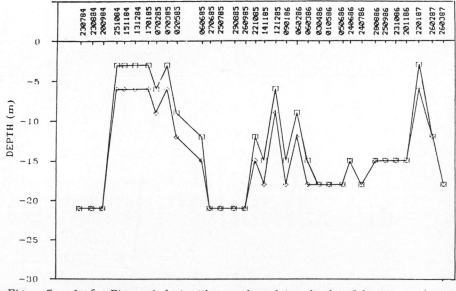

Figure 5.   As for Figure 4, but with sampling dates displaced by two weeks as
described in text.

Destratification of Hinze Dam was continued during the summer of 1985-86, but an oxycline is still evident until March 1986 when full aeration of the dam occurred. The presence of substantial DO throughout the dam profile was then retained by combination of aeration and winter "overturn" until January 1987. Problems associated with the dam extensions have prevented further work on the dam since then.

The efficacy of mixing via aeration is probably best seen in the success of reducing the temperature differential between surface and deep waters (Figure 6). This figure represents the difference in temperature between the surface and the 18 m depth for the dates shown. It can be clearly seen that the October 1985 aeration has an immediate mixing effect as shown in this temperature variation. While in midsummer of 1984-85 the temperature differential between surface and 18 m is greatest, by midsummer 1985-86 this has been markedly reduced. The concomitant raising of the temperature of the lower levels of the dam is not without its problems of increased algal growth in the hypolimnic regions, such growth leading to filtration problems in the treatment plant.

The amount of DO increase in the dam is not so easily assessed, however, there is a drop in differential between surface and deep waters in DO concentration due to an increase in DO at lower levels rather than a decrease at the surface (Figure 7). Again this figure represents the difference in concentration of (in this case) DO at the surface and at the 18 m level plotted against sampling date.

The effect upon manganese concentrations is very interesting (Figure 8). Each point in the figure gives a measure of average manganese concentration down to 15 m, obtained by averaging the manganese concentrations found at 0, 3, 6, 9, 12 and 15 m.

Reference to the period October 1984 to July 1985 prior to the destratification (Figure 8), indicates a normal behavioural pattern of Hinze Dam with regard to seasonal variations in manganese levels. By midspring, release of 0.45 µm filterable manganese

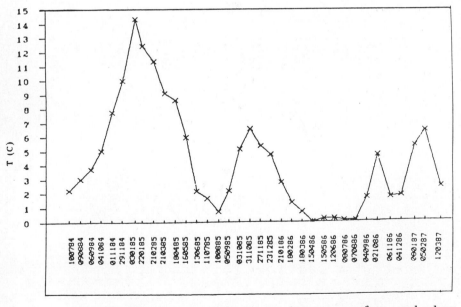

Figure 6. Plot of difference in temperature between surface and deep hypolimnion waters (18 m depth) over a three year period (Hinze Dam, 10-7-84 to 12-3-87).

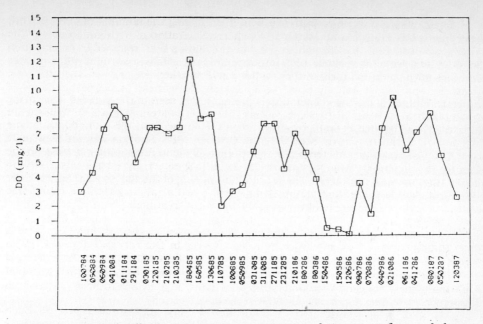

Figure 7.    Plot of difference in oxygen concentration between surface and deep hypolimnion waters (18 m depth) over a three-year period (Hinze Dam, 10-7-84 to 12-3-87).

from the sediment starts to increase the concentration of total manganese markedly, the maximum concentration being typically reached in late autumn or even early winter (June). At this later time the dam naturally destratifies and becomes oxic throughout the water column, with consequent rapid and marked depletion in manganese concentration at all levels. North Pine Dam has been shown to follow a very similar pattern of behaviour (Mokhtar, 1987).

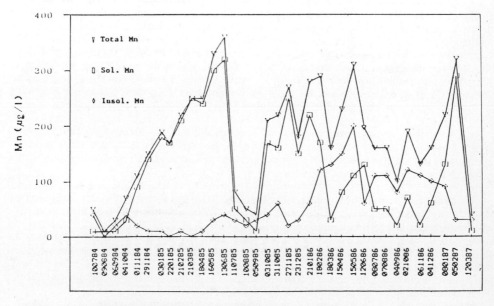

Figure 8.    Average manganese concentration in the water column (0 to 15 m) of Hinze Dam from 10-7-84 to 12-3-87.

149

When aeration of the dam commenced in October 1985, a rapid increase in soluble and total manganese was achieved in the dam waters. The rapidity of this phenomenon was probably due to stirring of the hypolimnic waters by the aeration process. It is significant that the 0.45 mm insoluble manganese now is a much greater percentage of the total manganese than was the case in the previous year when the total manganese was almost entirely made up of soluble manganese. A noticeable odour of hydrogen sulphide above the dam water during initial aeration indicated that base sediment had been disturbed into the water column; such sediments are rich in both soluble and insoluble forms of manganese.

To assess the effect of aeration upon the concentration of total soluble manganese, values for this parameter from Figure 8 for the three periods, October 1984 to June 1985, October 1985 to June 1986 (10 readings in each period) and October 1986 to March 1987 (6 readings) were averaged to yield the following:

| | |
|---|---|
| 1984-85 | 187 µg/L |
| 1985-86 | 122 µg/L |
| 1986-87 | 91 µg/L |

These estimates can be used to support the general conclusion that the total soluble manganese concentration was reduced by approximately 35% during the 1985-86 aeration period, and by approximately 50% in the 86-87 period.

If a similar assessment is undertaken for the concentrations of total manganese during the same three periods, the following results are obtained:

| | |
|---|---|
| 1984-85 | 217 µg/L |
| 1985-86 | 230 µg/L |
| 1986-87 | 170 µg/L |

Clearly the overall decrease of manganese in the water column is not marked, indeed aeration in the 1985-86 period may have put more manganese into the water column.

The fact that Figure 8 clearly shows that after aeration the concentration of insoluble manganese has increased by some orders of magnitude, indicates that there is clear change from soluble to insoluble manganese with destratification. However, if one wishes, as is normally the case, to draw water for treatment from just above the thermocline, or in a destratified dam from a depth of 9 m or above, there are problems. The average concentration of manganese in any form above the oxycline is much larger after destratification than before (Figure 9), while the total concentration of manganese down to the 9 m depth has not changed, although there is evidence that some soluble manganese has been converted to an insoluble form (Figure 10).

A further serious consequence of the results of this destratification/aeration attempt, which is clearly shown in Figure 8, is the marked fluctuation in concentration of manganese in all its forms during the aeration periods. Such fluctuation does not enhance the ease of raw water treatment.

Work on the insoluble manganese material (0.45 µm filtration) in the epilimnion of North Pine Dam has shown that, in many cases, the manganese in this form is readily resolubilized in dilute magnesium chloride solution at neutral pH (Mokhtar, 1987). In some cases up to 100% dissolves as $Mn^{2+}$(aq) ions under such treatment. Such work would appear to indicate that the insoluble manganese in the epilimnion of a stratified dam is not $MnO_2$, which would not be expected to dissolve by exchange reaction with $MgCl_2$ solution. It would be very interesting to see if destratification yields manganese oxides which do not resolubilize readily.

Unfortunately, only some three months of $MgCl_2$ exchange reactions on destratified Hinze Dam water insoluble manganese could be undertaken before dam

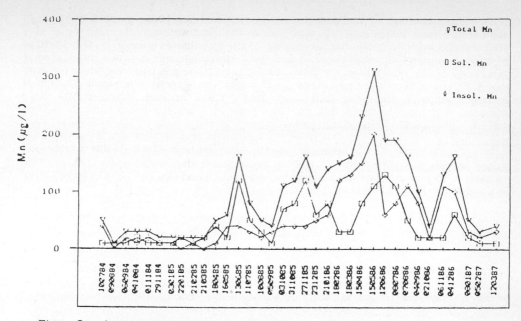

Figure 9.  Average manganese concentration in the water column above the oxycline of Hinze Dam from 10-7-84 to 12-3-87.

extensions prevented further work. The results, which will remain incomplete, indicate that little exchange occurred and it is suggested that destratification may yield oxidized manganese species. This phenomenon may be studied further in future if, and when, North Pine Dam is destratified.

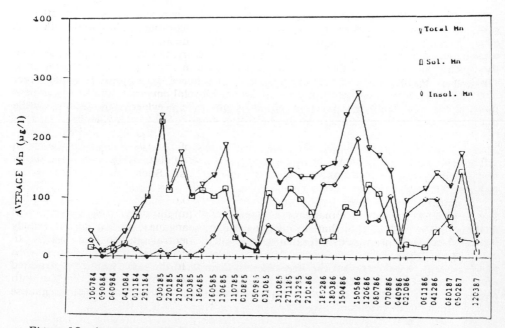

Figure 10.  Average manganese concentration in the water column down to 9 m for Hinze Dam from 10-7-84 to 12-3-87.

151

## CONCLUSIONS

Destratification of reservoirs by aeration has been used to obtain more consistent water quality for treatment to yield potable water. The technique has also been claimed to reduce the concentrations of iron and manganese present in the raw water of the dam, and consequently to reduce, or even to do away with, the need for a specific precipitation process during treatment to remove the metals.

Our studies on speciation of manganese in Hinze Dam (near Brisbane), both before and after destratification, have shown that the technique may, instead of removing the metal as a deposit to the sediments, change the nature of both insoluble and soluble forms of manganese. Therefore, under certain conditions more manganese may be present in the raw water after destratification than was present before aeration was commenced.

The need for a much better controlled and more detailed study of the processes accompanying destratification is highlighted.

## ACKNOWLEDGEMENTS

This work has been supported by a grant from the Queensland Department of Local Government. Acknowledgement of help in collecting samples and analyzing data is made to the Gold Coast City Council. The award of a scholarship (to M.B.M.) by the Public Services Department of Malaysia and the study leave grant (to M.B.M.) by the National University of Malaysia are gratefully appreciated.

## REFERENCES

Allaby, M., 1977. "A Dictionary of the Environment", The Macmillan Press, London.

Brown, I.K., Wolley, D.A. and Jory, A.G., 1982. Artificial destratification of Lake Morris to improve water quality, in: "Symposium on Hydrology and Water Resources", Melbourne 11-23 May, 1982, National Conf. Publications No. 82/3, The Institution of Engineers, Australia.

Burns, F.L., 1981. Experiences in the design, installation and operation of destratification aerators; in: "Destratification of Lakes and Reservoirs to Improve Water Quality", Aust. Water Resources Council, Conference Series No. 2. F.L. Burns and I.J. Powling, eds., Aust. Govt. Printing Service, Canberra.

Chiswell, B. and Rauchle, G., 1986. Manganese in water storage dams, *Proc. Roy. Soc. Queensland*, 92:53-67.

Davison, W. and Woof, C., 1984. A study of the Cycling of Manganese and Other Elements in a Seasonally Anoxic Lake, Rostherne Mere, U.K. *Water Res.*, 18:727-734.

Davison, W., Woof, C. and Rigg, E., 1982. The dynamics of iron and manganese in a seasonally anoxic lake; direct measurement of fluxes using sediment traps, *Limnol. Oceanogr.*, 27:987-1003.

Delphino, J.J. and Lee, G.F., 1968. Chemistry of manganese in Lake Mendota, Wisconsin. *Environ. Sci. Technol.*, 2:1094-1100.

Hutchinson, G.E., 1957. "A Treatise on Limnology", Vol. 1, John Wiley, New York.

McMahon, J.W., 1969. The annual and diurnal variation in the vertical distribution of acid-soluble ferrous and total iron in a small dimictic lake, *Limnol. Oceanogr.*, 14:357-367.

Mokhtar, M.B., 1987. "Manganese Speciation In Dam Waters", PhD Thesis, University of Queensland, St. Lucia, Australia.

Nyffler, U.P., Schindler, P.W., Wirz, U.E. and Imboden, D.M., 1983. Chemical and geochemical studies of Lake Biel, *Schweiz. Z. Hydrol.*, 45:48-49.

Welsh, P.R., 1984. "An Assessment of Water Quality and Monitoring at Dartmouth Reservoir", Rural Water Commission of Victoria, Report No. 82, Melbourne.

Wetzel, R.G., 1975. "Limnology", Saunders College Publishing, Washington.

## CONTRIBUTORS

Dr Ronald C. Bayly

Department of Microbiology
Monash University
Wellington Road
CLAYTON   VICTORIA  3168


Dr Ron Beckett

Water Studies Centre
Department of Chemistry
Monash University
PO Box 197
CAULFIELD EAST   VICTORIA  3145


Dr Brian A. Bolto

Division of Chemicals & Polymers
Commonwealth Scientific & Industrial Research Organization
Bayview Avenue
CLAYTON   VICTORIA  3168


Dr Barry Chiswell

Department of Chemistry
University of Queensland
BRISBANE   QUEENSLAND  4000


Dr David R. Dixon

Division of Chemicals & Polymers
Commonwealth Scientific & Industrial Research Organization
Bayview Avenue
CLAYTON   VICTORIA  3168


Dr Annabelle Duncan

Department of Microbiology
Monash University
Wellington Road
CLAYTON   VICTORIA  3168


Dr Brian L. Finlayson

Centre for Environmental Applied Hydrology
Department of Geography
University Melbourne
PARKVILLE   VICTORIA  3052


Prof. Thomas W. Healy

Department of Physical Chemistry
University of Melbourne
PARKVILLE   VICTORIA  3052


Dr Keith A. Hunter

Department of Chemistry
University of Otago
DUNEDIN   NEW ZEALAND

| | |
|---|---|
| Peter R. Hutchison | Iron Chemicals Division<br>Tioxide Australia Pty Ltd<br>PO Box 184<br>BURNIE   TASMANIA  7320 |
| Dr Luis O. Kolarik | Division of Chemicals & Polymers<br>Commonwealth Scientific & Industrial Research Organization<br>Bayview Avenue<br>CLAYTON   VICTORIA  3168 |
| Dr Richard T. Lowson | Australian Nuclear Science & Technology Organisation<br>Private Mail Bag 1<br>MENAI   NEW SOUTH WALES  2234 |
| Prof. Kevin C. Marshall | School of Microbiology<br>University of New South Wales<br>KENSINGTON   NEW SOUTH WALES 2033 |
| Dr John W. May | Department of Microbiology<br>Monash University<br>Wellington Road<br>CLAYTON   VICTORIA  3168 |
| Dr Mazlin Bin Mokhtar | Chemistry Department<br>University of Queensland<br>BRISBANE   QUEENSLAND   4000 |
| Dr William G.C. Raper | Division of Chemicals and Polymers<br>Commonwealth Scientific & Industrial Research Organization<br>Bayview Avenue<br>CLAYTON   VICTORIA  3168 |
| Dr Stephen A. Short | Australian Nuclear Science & Technology Organisation<br>Private Mail Bag 1<br>MENAI   NEW SOUTH WALES  2234 |
| Dr George Vasiliadis | Department of Microbiology<br>Monash University<br>Wellington Road<br>CLAYTON   VICTORIA  3168 |
| Dr T. David Waite | Australian Nuclear Science & Technology Organisation<br>Private Mail Bag 1<br>MENAI   NEW SOUTH WALES  2234 |

# INDEX

Coagulation domains, 127-129
Coastal, 11
Co-ion, 122
Colligative properties, 4
Collision, 46-48, 132, 133
Colloid, 3, 9, 10-12, 15, 21, 25, 45, 47, 49, 65, 71-75, 78-83, 92, 103, 117, 119-127, 132
Colloid aggregation, 48
Colloid stability, 9, 12, 16, 22, 48, 121, 123, 125
Colour, 4, 88-95, 99, 103-106, 108, 109, 113, 116, 119, 122, 123, 129, 132
Complexation, 11, 13, 32, 79, 83, 112, 123
Complexes, 12, 13, 36, 106-108, 120, 123, 130-132
Complexing capacity, 4
Condensation, 5
Conduction band, 30, 31, 35, 40
Conductivity, 79, 106
Continuous flow centrifuge, 65
Coordination, 12, 129-131
Copiotrophic bacteria, 24
Copper, 12-14, 40, 41, 97, 105
Counter-ion, 122, 126
Critical coagulation concentration, 125-127
Critical stabilization concentration, 126, 127
Cyanide, 13, 14, 39

Dam, 143-152
Denitrification, 98, 135, 136
Desalination, 89, 94, 95
Desorption, 9, 31, 106, 115
Destabilization, 119-133
Destratification, 143-152
Dialysis, 24, 25, 71, 94, 95, 97
Diffusion, 12, 31, 74
Disinfection, 88, 94, 98, 99, 119, 123
Dissolution, 9, 11, 27-33, 41, 90, 144
Dissolved organic carbon, 3, 4, 7, 12, 13
Dissolved organic matter, 3, 15
Distillation, 89, 94, 95
DLVO Theory 22, 123
DOC, *see* Dissolved organic carbon
DOM, *see* Dissolved organic matter
Drinking water, 88, 98, 143

Effluent, 39, 40, 96, 98, 99, 103, 123, 135, 136, 138
Electrical double layer, 22, 48, 122-126, 132
Electrocoagulation, 93, 94
Electrodialysis, 95, 97
Electrokinetic properties, 21, 107, 112, 113

Electron-hole pair, 30, 35
Electrophoresis, 9, 21, 122
Electrophoretic mobility, 9, 10, 122, 127, 128
Electrostatically stabilized colloid, 48
Endogenous respiration, 24
Epilimnion, 143, 150
Erosion, 57, 59, 61, 67
Estuary, 10, 11, 45-53, 106
Ethanol, 97, 98
Eutrophication, 93, 135, 136, 140
Extracellular polymer, 21
Extraction, 15, 98, 136

Faecal coliform, 98
Ferric hydroxide, 14, 81
Ferric oxide, 33
Ferrihydrite, 81
Fertilizer, 88, 97
Fibril, 22, 23
Field-flow fractionation, 4
Filter, 30, 45-48, 65, 90-94, 99, 133
Filterable component, 12, 45, 48, 51, 144, 148
Filtration, 45, 89-93, 99, 124, 133, 144, 148, 150
Fish, 13, 67, 92, 94
Floc, 89-94, 124, 126, 132, 133, 136
Flocculation, 12, 89, 90, 92-94, 96, 119, 120, 125, 132, 133
Flotation, 8, 90, 91, 93, 97
Flow velocity, 65
Fluoride, 52, 87, 88, 94, 98
Foam, 8, 16
Food chain, 45
Freshwater, 10, 11, 13, 14, 45, 49-52, 99, 144
Fulvic acid, 4, 6, 9, 10, 30, 33, 111, 112

Gel filtration, 71
Gibbs adsorption equation, 6
Glyphosate, 15
Goethite, 9, 10, 13, 14, 15, 32
Gold, 39, 143
Grain size, 65, 67
Gram-negative bacteria, 22
Gram-positive bacteria, 22
Ground water, 4, 11, 71-83, 94

Haloforms, 36, 123
Halogenated aliphatics, 36, 37, 123
Hardness, 9, 11, 87, 88, 94, 97, 105-108, 112
Heavy metals, 40, 93, 94, 97, 98, 105, 108, 109
Hematite, 12, 20, 31, 32, 35
Herbicides, 15, 27, 28, 37, 41, 105, 110, 112
Heterocoagulation, 105, 106, 112